T0122327

Recent advances in statistics have led to new concepts and solutions in different areas of pharmaceutical research and development. "Springer Series in Pharmaceutical Statistics" focuses on developments in pharmaceutical statistics and their practical applications in the industry. The main target groups are researchers in the pharmaceutical industry, regulatory agencies and members of the academic community working in the field of pharmaceutical statistics. In order to encourage exchanges between experts in the pharma industry working on the same problems from different perspectives, an additional goal of the series is to provide reference material for non-statisticians. The volumes will include the results of recent research conducted by the authors and adhere to high scholarly standards. Volumes focusing on software implementation (e.g. in SAS or R) are especially welcome. The book series covers different aspects of pharmaceutical research, such as drug discovery, development and production.

More information about this series at http://www.springer.com/series/15122

Jozef Nauta

Statistics in Clinical and Observational Vaccine Studies

Second Edition

Jozef Nauta
Amsterdam, The Netherlands

ISSN 2366-8695 ISSN 2366-8709 (electronic)
Springer Series in Pharmaceutical Statistics
ISBN 978-3-030-37695-6 ISBN 978-3-030-37693-2 (eBook)
https://doi.org/10.1007/978-3-030-37693-2

For copies of the SAS codes, please write an e-mail to jozef.nauta@gmail.com.

This Springer imprint is published by the registered company Springer Nature Switzerland AG
The registered company address is: Gewerbestrasse 11, 6330 Cham, Switzerland

It was about the beginning of September, 1664, that I, among the rest of my neighbours, heard in ordinary discourse that the plague was returned again in Holland; for it had been very violent there, and particularly at Amsterdam and Rotterdam, in the year 1663, whither, they say, it was brought (some said from Italy, others from the Levant) among some goods which were brought home by their Turkey fleet; others said it was brought from Candia; others, from Cyprus. It mattered not from whence it came; but all agreed it was come into Holland again.

Daniel Defoe
A Journal of the Plague Year (1722)

Preface

In the ten years between the publication of the first edition of this book and that of this second edition, the world witnessed two major virus outbreaks. The first was the West African Ebola outbreak, the most widest outbreak of Ebola virus disease in history. At the time of the outbreak, no vaccine was available. A large vaccine efficacy study with an experimental Ebola vaccine involving 11,841 people was conducted in Guinea during 2015 [1]. The vaccine was highly protective. Among the 5,837 participants who received the vaccine, no Ebola cases occurred. In comparison, among the 6,004 participants who had not received the vaccine 23 cases occurred. The second outbreak was that of the Zika virus in South America, around the time of the 2016 Olympic Summer Games in Brazil. The United States National Institute of Allergy and Infectious Diseases is developing multiple vaccine candidates to prevent Zika virus infection. Vaccine research and development continues to play a major role in greatly reducing disease, disability and death worldwide.

This book is intended for statisticians working in clinical vaccine development in the pharmaceutical industry, at universities, at national vaccines institutes, etc. Statisticians already involved in clinical or observational vaccine studies may find some interesting new ideas in it, while colleagues who are new to vaccine development or vaccine epidemiology will be able to familiarize themselves quickly with the statistical methodology.

A good knowledge of statistics is assumed. The reader should be familiar with hypothesis testing, point and confidence interval estimation, likelihood methods, bootstrapping, etc. Nonetheless, the scope of the book is practical rather than theoretical. Many real-life examples are given, and SAS codes are provided, making application of the methods straightforward.

The book is divided into five parts. Part I comprises two chapters that should be read in tandem, and which will provide the reader with the necessary background knowledge of the fundamentals of vaccination and the working of the immune system. Part II is dedicated to the analysis of immunogenicity data. New to this second edition are the four chapters in Part III on vaccine field studies, i.e. vaccine efficacy and vaccine effectiveness studies. The first chapter, Chap. 7, serves as an introduction to vaccine field studies. In Chap. 8, the analysis of vaccine efficacy data and the assumptions underlying the different analyses are discussed. In Chap. 9, observational vaccine studies are explored. (Hence the change of the title of the book.) The meta-analysis of vaccine effectiveness studies is the topic of Chap. 10. Part IV addresses correlates of protection, immunological assays that predict protection against infection. In Part V, the analysis of vaccine safety data is discussed.

Also new are Appendices E–I. In Appendix E, a proof of the 'exponential formula' is presented, the formula that links the occurrence measures risk of infection and force of infection. In Appendix F, it is shown that although a force of infection function and a hazard function are conceptually different, they take the same functional form and that in this sense they are synonyms. In Appendix G, it is explained how a confidence interval for the difference of two vaccine effectiveness estimates can be obtained by means of bootstrapping. Appendix H contains a SAS code for the creation of the data of the examples of Chap. 8. In Appendix I, some more SAS codes are presented, including a code for Barnard's exact test for rate ratios and a code for a bootstrap approach to compare two risk estimates based on the popular Nelson–Aalen estimator.

Several changes were made to the existing chapters. Chapter 11 on correlates of protection has been thoroughly revised. Several examples were rewritten. Last but not least, the notation and the terminology has been improved.

April 2018

Jozef Nauta
Museo Galileo, Florence, Italy

Acknowledgement

I would like to acknowledge my colleague and good friend Dr. Walter Beyer of the Department of Virology, Erasmus Medical Centre, Rotterdam, the Netherlands, for his generous advice on Chaps. 1, 2 and 10, and the many inspiring discussions we had while working on our joint publications.

Jozef Nauta

Contents

About the Author

Jozef Nauta is a principal statistician, with more than 25 years of experience in the pharmaceutical industry, and with special interest in the development of influenza vaccines. During his career, he has published numerous journal articles on statistics and vaccines. He currently participates in DRIVE (Development of Robust and Innovative Vaccine Effectiveness), a public–private partnership that advances European cooperation in influenza vaccine effectiveness studies, and funded by, amongst others, IMI (Innovative Medicines Initiative). He lives in Amsterdam, The Netherlands, together with his wife and son.

Acronyms

AAP	American Academy of Pediatrics
AIDS	Acquired immune deficiency syndrome
ANCOVA	Analysis of covariance
AOM	Acute otitis media
ARI	Acute respiratory illness
BCG	Bacillus Calmette–Guérin
CBER	Center for Biologics Evaluation and Research
CDC	Centers for Disease Control and Prevention
CF	Cystic fibrosis
CGD	Chronic granulomatous disease
CL	Confidence limit
CLRS	Constrained likelihood ratio statistic
CMI	Cell-mediated immunity
CoP	Correlate of protection
CTL	Cytotoxic T lymphocyte
DAG	Directed acyclic graph
DNA	Deoxyribonucleic acid
DTP	Diphtheria, tetanus, pertussis
EIA	Enzyme immunoassay
ELISA	Enzyme-linked immunosorbent assay
ELISPOT	Enzyme-linked immunospot
EMA	European Medicines Agency
EPPT	Events-per-person-time
FDA	United States Food and Drug Administration
FDR	False discovery rate
FWER	Family-wise error rate
GBS	Guillain Barré syndrome
GM	Geometric mean
GP	Glycoprotein-based; General practitioner
HA	Haemagglutinin

HAI	Haemagglutination inhibition
HI	Haemagglutination inhibition
Hib	*Haemophilus influenzae* type b
HIV	Human immunodeficiency virus
HPV	Human papillomavirus
ICH	International Conference on Harmonisation of Technical Requirements for Registration of Pharmaceuticals for Human Use
IFN-γ	Interferon-gamma
i.i.d.	Independent and identically distributed
IL	Interleukin
ILI	Influenza-like illness
IM	Immune marker
IU	Intersection-union
LL	Log-likelihood
LRS	Likelihood ratio statistic
mCoP	Mechanistic correlate of protection
MedDRA	Medical Dictionary for Regulatory Activities
ML	Maximum likelihood
MMR	Measles, mumps, rubella
MMRV	Measles, mumps, rubella, varicella
nCoP	Non-mechanistic correlate of protection
NRA	National Registration Authority (Australia)
NIH	National Institutes of Health
NK	Natural killer
NMPA	National Medical Products Administration (China)
OPSR	Organization for Pharmaceutical Safety and Research (Japan)
PCR	Polymerase chain reaction
PBMC	Peripheral blood mononuclear cells
RCD	Reverse cumulative distribution
RNA	Ribonucleic acid
SBA	Serum bactericidal assay
SD	Standard deviation
SIDS	Sudden infant death syndrome
SPC	Spot-forming cell
TH1	T helper 1
TH2	T helper 2
TND	Test-negative design
TOST	Two one-sided tests
UTI	Urinary tract infection
V	Varicella
VAERS	Vaccine Adverse Event Reporting System
VLP	Virus-like particles

Notation and Terminology

In this book, population parameters and parameters of distributions are notated by Greek symbols, estimators and estimates of parameters by (combinations of) italic capitals. Random variables are notated as bold capitals.

μ	Mean of a normal (Gaussian) distribution
σ	Standard deviation of a normal distribution
$\Delta\,(=\mu_1-\mu_0)$	Difference of two normal means
e^{μ}	Median of a log-normal distribution
$\theta\,(=e^{\mu_1}/e^{\mu_0})$	Ratio of the medians of two log-normal distributions
β_i	Coefficient of a regression model
π	Risk
$\Delta\,(=\pi_1-\pi_0)$	Risk difference
$\theta\,(=\pi_1/\pi_0)$	Relative risk
λ	Force of infection
$\lambda\,(t)$	Force of infection function
$\Lambda\,(t)$	Cumulative force of infection function
$\phi\,(=\lambda_1/\lambda_0)$	Relative force of infection
$\vartheta\,(=1-\theta)$	Vaccine efficacy; vaccine effectiveness
$\vartheta_{\phi}\,(=1-\phi)$	Vaccine efficacy (approximation); vaccine effectiveness (approximation)
GMC	Geometric mean concentration, estimator/estimate (*est.*) of the median concentration
GMFR	Geometric mean fold ratio, *est.* of the ratio of two median fold increases
GMR	Geometric mean ratio, *est.* of the ratio of two median titres
GMT	Geometric mean titre, *est.* of a median titre
GSD	Geometric standard deviation, antilog of the sample standard deviation of a set of log-transformed immunogenicity values
B_i	Estimate of a regression coefficient
R	Rate, *est.* of a risk

RD	Rate difference, *est.* of a risk difference
RR	Rate ratio, *est.* of a relative risk
RRE	Relative risk estimate in a meta-analysis of vaccine effectiveness estimates
IR	Incidence rate, *est.* of homogeneous force of infection
IRR	Incidence rate ratio, *est.* of a relative force of infection
OR	Odds ratio, *est.* of a relative risk of infection or a relative force of infection
HR	Hazard rate, *est.* of a hazard ratio
VE	*Est.* of the vaccine efficacy or the vaccine effectiveness
SD(..)	Sample standard deviation
SE(..)	Standard error of an estimator, either asymptotic or exact
LCL	Lower confidence limit
UCL	Upper confidence limit
M	Equivalence/non-inferiority margin

For estimators and estimates, the same notation is used. An *estimator* is a function, for example:

$$IRR = \frac{c_1/T_1}{c_0/T_0}.$$

When the function is applied, an *estimate* of a parameter is obtained, for example:

$$IRR = 0.328.$$

Occasionally, estimators and estimates are notated by a 'hooded' parameter: $\hat{\mu}$, $\hat{\theta}$, etc.

The term *rate* is usually reserved for estimators of the form number of cases per unit of time. In this book, the term is used in what may be called the proportion sense, for estimators of the form proportion of cases that occurred in a fixed group of individuals. The justification is that in clinical vaccine trials, there are a number of standard concepts that, although proportions, are called rates: seroprotection rate, seroconversion rate and attack rate. For estimators of the form cases per unit of time, the term *incidence rate* is used.

Part I
The Interplay Between Microorganisms and the Immune System

Chapter 1
Basic Concepts of Vaccine Immunology

Abstract The first two chapters of this book are intended to provide the reader with the necessary background knowledge of the fundamentals of vaccination. This chapter opens with an overview of the major infectious microorganisms. Next, the working of the immune system is explained, how it can ward off microorganisms it has encountered before. The primary defence mechanism of microorganisms—antigenic variation—is examined. An overview of the several types of vaccines for viruses and bacteria, from the first-generation live attenuated vaccines to third-generation vaccines such as recombinant vector vaccines, DNA vaccines and virus-like particles vaccines is given.

1.1 Vaccination and Preventing Infectious Diseases

Vaccines take advantage of the body's ability to learn how to ward off microorganisms. The immune system can recognize and fight of quickly infectious organisms it has encountered before. As an example, consider chickenpox. Chickenpox is a highly contagious infectious disease caused by the varicella-zoster virus. First, there are papules, pink or red bumps. These bumps turn into vesicles, fluid-filled blisters. Finally, the vesicles crust over and scab. Clinical symptoms are fever, abdominal pain or loss of appetite, headache, malaise and dry cough. The disease is so contagious that most people get it during their childhood, but those infected are the rest of their life immune to it. Vaccines contain killed or inactivated (parts of) microorganisms. These provoke the immune system in a way that closely mimics the natural immune response to the microorganisms. Vaccination is a less risky way to become immune, because, due to the killing or inactivation of the microorganisms, it does not cause the disease.

Vaccination, together with hygiene, is considered to be the most effective method of preventing infectious diseases. When not prevented, some infectious diseases have proven to be mass killers. Plague, caused by the bacterium *Yersinia pestis*, has been one of the deadliest pandemics in history. The total number of plague deaths worldwide has been estimated at 75 million people, and the disease is thought to have killed almost half of Europe's population. The pandemic arrived in Europe in

J. Nauta, *Statistics in Clinical and Observational Vaccine Studies*, Springer Series in Pharmaceutical Statistics, https://doi.org/10.1007/978-3-030-37693-2_1

the fourteenth century, and it would cast its shadow on the continent for five centuries, with one of the last big outbreaks occurring in Moscow in 1771. (The reader who wants to learn how it was to be trapped in a plague-stricken community should read Giovanni Boccaccio's *Il Decameron* (1353), Daniel Defoe's *A Journal of the Plague Year* (1722) or Albert Camus' *La Peste* (1947).)

The global death toll from the Spanish influenza pandemic (1918–1920), caused by an influenza virus, is assumed to have been 50 to 100 million people, more than the combined total casualties of World Wars I and II.

Malaria is a potentially deadly tropical disease transmitted by a female mosquito when it feeds on blood for her eggs. In Africa, an estimated 2,000 children a day die from the disease, leading in 2006 to a total number of deaths from the disease of almost one million. The Bill and Melinda Gates Foundation is funding efforts to reduce malaria deaths, by developing more effective vaccines. The long-term goal of the foundation is to eradicate the disease.

1.2 Microorganisms: Bacteria, Yeasts, Protozoa and Viruses

Microorganisms (also microbes) are live forms that cannot be seen by the unaided eye, but only by using a light or an electron microscope. The Dutch scientist Anton van Leeuwenhoek (1632–1723) was the first to look at microorganisms through his microscope. Microorganisms that cause disease in a host organism are called *pathogens*. If a microorganism forms a symbiotic relationship with a host organism of a different species and benefits at the expense of that host, it is called a *parasite*.

Bacteria are unicellular organisms surrounded by a cell wall and typically 1–5 μm in length. They have different shapes such as rods, spheres and spirals, and reproduce asexually by simple cell division. The biological branch concerned with the study of bacteria is called bacteriology. Examples of serious bacterial diseases are diphtheria, tetanus, pertussis, cholera, pneumococcal disease, tuberculosis, leprosy and syphilis.

Yeasts are unicellular organisms typically larger than bacteria and measuring around 5 μm. Most reproduce asexually, but some also show sexual reproduction under certain conditions. Yeasts are studied within the branch of mycology. Diseases caused by yeasts are, among others, thrush and cryptococcosis.

Protozoa are unicellular organisms, more complex and larger than bacteria and yeasts, typically between 10 and 50 μm in diameter. They usually are hermaphroditic and can reproduce both sexually and asexually. Protozoa are responsible for widespread tropical diseases such as malaria, amoebiasis, sleeping sickness and leishmaniasis. The biological branch of parasitology includes the study of protozoa and of certain multicellular organisms such as Schistosoma and helminths (parasitic worms).

In contrast with bacteria, yeasts and protozoa, which are cellular live forms, *viruses* are too small to form cells (typically 0.05–0.20 μm in diameter). In the environment, they show no metabolism. For replication, a virus needs to intrude a host cell and take over the cell metabolism to produce and release new virus particles. Viruses contain either DNA or RNA as genetic material. DNA viruses include herpes-, adeno-, papova-, hepadna- and poxviruses. RNA viruses include rhino-, polio-, influenza- and rhabdoviruses. Some RNA viruses have an enzyme called reverse transcriptase that allows their viral RNA to be copied as a DNA version (retroviruses). Well-known viral diseases are herpes, hepatitis B and smallpox (DNA viruses), common cold, poliomyelitis, hepatitis A, influenza, rabies (RNA viruses) and human immunodeficiency virus (HIV) (RNA retroviruses). The study of viruses is called *virology*.

1.3 The Immune System

1.3.1 Basics

The immune system can distinguish between non-foreign and foreign (also self and non-self) molecules and structures. With this ability, it seeks to protect the organism from invading pathogens—by detecting and killing them. The immune system has two essential components, the innate (inborn) or non-specific and the adaptive or specific immune system.

The *innate immune system* provides an immediate, albeit non-specific, response to invading pathogens. It is triggered by cells and molecules that recognize certain molecular structures of microorganisms, and it tries to inhibit or control their replication and spread. In vertebrates, one of the first responses of the innate immunity to infection is inflammation, initiated by infected and injured cells that, in response, release certain molecules (histamine, prostaglandins and others). These molecules sensitize pain receptors, widen local blood vessels and attract certain white blood cells (neutrophils) circulating in the bloodstream and capable to kill pathogens by ingestion (*phagocytosis*) as a front-line defence. Neutrophils can release even more signalling molecules such as *chemokines* and *cytokines* (among many others: interferon-γ) to recruit other immune cells, including macrophages and natural killer cells. *Macrophages* reside in tissue and also ingest and destroy pathogens. *Natural killer* (NK) cells can detect infected cells (and some tumour cells) and destroy them by a mechanism which is known as *apoptosis*, cell death characterized by protein and DNA degradation and disintegration of the cell. The innate immune system responds to microorganisms in a general way during the early phase of the infection, and it does not confer long-lasting immunity. In vertebrates, the innate immune system actives the adaptive immune system in case pathogens successfully evade this first line of defence.

The *adaptive immune system* has the remarkable ability to improve the recognition of a pathogen, to tailor a response specific to the actual structure of that pathogen, and to memorize that response as preparation for future challenges with the same or a closely similar pathogen. The adaptive immune system activates bone marrow-derived (*B cells*) and thymus-derived cells (*T cells*), leading to humoral and cellular immunity, respectively (see also Chap. 2). In general, B cells make antibodies that attack the pathogens directly, while T cells attack body cells that have been infected by microorganisms or have become cancerous. When activated, B cells secrete *antibodies* in response to *antigens* (from antibody-generating), molecules recognized as non-self. An antigen can be a part of a microorganism, a cancerous structure or a bacterial toxin. The antibodies that are produced are specific to that given antigen. The major role of antibodies is either to mark the invaders for destruction (which, in turn, is effected by other immune cells) or to inactivate (neutralize) them so that they can no longer replicate.

Like B cells, T cells have surface receptors for antigens. T cells can specialize to one of several functions: They may help B cells to secrete antibodies (*T helper cells*), attract and activate macrophages, or destroy infected cells directly (*cytotoxic T cells*, also *T killer cells*). This improved response is retained after the pathogen has been killed (*immunological memory*). It allows the immune system to react faster the next time the pathogen invades the body. This ability is maintained by memory cells which remember specific features of the pathogen encountered and can mount a strong response if that pathogen is detected again.

In vertebrates, the *immune system* is a complex of organs, tissues and cells connected by two separate circulatory systems, the bloodstream and the lymphatic system that transports a watery clear fluid called lymph.

In the red bone marrow, a tissue found in the hollow interior of bones, multipotent stem cells differentiate to either red blood cells (erythrocytes), or platelets (thrombocytes), or white blood cells (leukocytes). The latter class is immunologically relevant; leukocytes maturate to either granulocytes (cells with certain granules in their cytoplasm and a multi-lobed nucleus, for example, the neutrophils mentioned previously) or mononuclear leukocytes, including macrophages and lymphocytes. Natural killer cells, B cells and T cells belong to the lymphocytes. T cell progenitors migrate to the thymus gland, located in the upper chest, where they mature to functional T cells. In the spleen, an organ located in the left abdomen, immune cells are stored and antibody-coated microorganisms circulating in the bloodstream are removed. Finally, the lymph nodes store, proliferate and distribute lymphocytes via the lymphatic vessels.

1.3.2 Microbial Clearance

Virus clearance or elimination of a virus infection involves killing of infected cells by NK cells and cytotoxic T cells, blocking of cell entry or cell-to-cell transmission by neutralizing antibodies, and phagocytosis by macrophages.

The major process of bacterial clearance is phagocytosis. Pathogenic bacteria have three means of defence against it. The first defence is the cell capsule, a layer outside the cell wall that protects bacteria from contact with macrophages and other phagocytes (cells that protect the body by ingesting harmful foreign particles and dead or dying cells). The second defence is the cell wall, which acts as a barrier to microbicidal activity. The third defence is the secretion of *exotoxins*, poisonous substances that damage phagocytes and local tissues and, once circulating in the bloodstream, remote organs. Frequently, exotoxins (and not the bacteria themselves) are the cause of serious morbidity of an infected organism. Most cell capsules and exotoxins are antigenic, meaning that antibodies can block their effects.

Protozoan clearance is exceptionally difficult. Immunity is usually limited to keeping the parasite density down. Malaria clearance, for example, involves phagocytosis of parasitized red blood cells by macrophages and antibodies. During the brief liver stage of the malaria parasites, immunity can be induced by cytotoxic T cells.

1.3.3 Active and Passive Protection from Infectious Diseases

The immune system can quickly recognize and fight off infectious organisms it has encountered before. Measles is a highly contagious infectious childhood disease caused by the measles virus and transmitted via the respiratory route. Infected children become immune to it for the rest of their life. This is called *naturally acquired active immunity*. Because newborn infants are immunologically naive (no prior exposure to microorganisms), they would be particularly vulnerable to infection. Fortunately, during pregnancy, antibodies are passively transferred across the placenta from mother to foetus (*maternal immunity*). This type of immunity is called *naturally acquired passive immunity*. Depending on the half-life time of these passively transferred antibodies, maternal immunity is usually short-term, lasting from a few days up to several months.

1.3.4 Antigenic Variation

While measles does usually not infect an individual twice in lifetime due to naturally acquired active immunity, some other pathogens try to trick the immunological memory by various mechanisms. One is an adaptation process called antigenic variation: small alterations of the molecular composition of antigens of the surface of microorganisms to become immunologically distinct from the original strain. (A *strain* is a subset of a species differing from other members of the same species by some minor but identifiable change.) *Antigenic variation* can occur either due to gene mutation, gene recombination or gene switching. Antigenic variation can occur very slowly or very rapidly. For example, the poliovirus, the measles virus and the yellow fever virus have not changed significantly since vaccines against them were first developed, and these vaccines therefore offer lifelong protection. Examples for rapidly evolving viruses are HIV and the influenza virus. Rapid antigenic variation is an important cause of vaccine failure.

A *serotype* is a variant of a microorganism in which the antigenic variations are to such a degree that it is no longer detected by antibodies directed to other members of that microorganism. For example, of the bacterium *Pseudomonas aeruginosa*, more than sixteen serotypes are known, of the hepatitis B virus four major serotypes have been identified, and of the rhinovirus, cause of the common cold, there are so many serotypes (more than 100) that many people suffer from common cold several times every winter—each time caused by a member of a different serotype. In case of influenza, antigenic variation is called *antigenic drift*, which is the process of mutations in the virus surface proteins haemagglutinin and neuraminidase. This drift is so rapid that the composition of influenza vaccines must be changed almost every year. Antigenic drift should not be confused with *antigenic shift*, the process at which two different strains of an influenza virus combine to form a new antigenic subtype, for which the immune system of the host population is naive and which makes it extremely dangerous because it can lead to pandemic outbreaks.

1.4 Vaccines

The word vaccination (Latin: *vacca*–cow) was first used by the British physician Edward Jenner (1749–1823) who searched for a prevention of smallpox, a widespread disease localized in small blood vessels of the skin, mouth and throat, causing a maculopapular rash and fluid-filled blisters and often resulting in disfigurement, blindness and death. In 1798, Jenner published his *An Inquiry into the Causes and Effects of the Variolae Vaccinae, a Disease Discovered in Some of the Western Counties of England, Particularly Gloucestershire, and Known by the Name of the Cow-Pox*. He reported how he, two years earlier, had taken the fluid from a cowpox pustule on a dairymaid's hand and inoculated an eight-year-old boy. Six weeks later, he exposed the boy to smallpox, but the boy did not develop any symptoms of smallpox disease. Today, the virological background of Jenner's successful intervention is understood: smallpox virus, the cause of smallpox, and cowpox virus, the cause of a mild veterinary disease with only innocent symptoms in men, are quite similar DNA viruses belonging to the same viral genus orthopoxvirus. Unintendedly, dairymaids were often exposed to and infected by cowpox virus during milking. Consequently, they developed immunity which also protected against the smallpox virus (cross-protection). Previously, this type of immunity was called naturally acquired active immunity. By intended inoculation with cowpox virus, Jenner had the eight-year-old boy actually achieve artificially acquired active immunity—the aim of any vaccination. The year 1996 marked the two hundredth anniversary of Jenner's experiment. After large-scale vaccination campaigns throughout the nineteenth and twentieth century using vaccinia virus, another member of the same viral genus, the World Health Organization in 1979 certified the eradication of smallpox. To this day, smallpox is the only human infectious disease that has been completely eradicated.

Among the pioneers of vaccinology were the French chemist Louis Pasteur (1822–1895), who developed a vaccine for rabies, and the German Heinrich

Hermann Robert Koch (1843–1910), who isolated *Bacillus anthracis*, *Vibrio cholerae* and *Mycobacterium tuberculosis*, a discovery for which he in 1905 was awarded the Nobel Prize in Physiology or Medicine. Koch also developed criteria to establish, or refute, the causative relationship between a given microorganism and a given disease (*Koch's postulates*). This was, and is, essential for vaccine development. First, one has to prove that a given microorganism is really the cause of a given clinical disease, and then one can include that microorganism in a vaccine to protect people from that disease. The causative relationship between a microbe and a disease is not always self-evident. In the first decades of the twentieth century, it was widely believed that the cause of influenza was the bacterium *Haemophilus influenzae*, because it was often isolated during influenza epidemics. Only when in the 1930s influenza viruses were discovered and proven, by Koch's postulates, to be the real cause of influenza, the way was opened to develop effective vaccines against that disease. A vaccine-containing *H. influenzae* would not at all protect from influenza.

Most vaccines contain *attenuated* (weakened) or inactivated microorganisms. Ideally, they provoke the adaptive immune system in a way that closely mimics the immune response to the natural pathogenic microorganisms. Vaccination is a less risky way to become immune, because, due to the attenuation or inactivation of the microorganisms, a vaccine does not cause the disease associated with the natural microorganism. Yet, naive B and T cells are activated as if an infection had occurred, leading to long-lived memory cells, which come into action after eventual exposure with the natural microorganism.

1.4.1 Viral and Bacterial Vaccines Currently in Use

Live attenuated vaccines contain living viruses or bacteria of which the genetic material has been altered so they cannot cause disease. The classical way of attenuation is achieved by growing the microorganisms over and over again under special laboratory conditions. This *passaging* process deteriorates the disease-causing ability of the microorganisms. The weakened viruses and bacteria still can infect the host, and thus stimulate an immune response, but they can rarely cause disease. However, in certain immune-compromised patients, even attenuated microorganisms may be dangerous so that manifest immune-suppression can be a contra-indication for live vaccines.

An example of a live attenuated vaccine is the RIX4414 human rotavirus vaccine. Rotavirus infection is the leading cause of potentially fatal dehydrating diarrhoea in children. The parent strain RIX4414 was isolated from a stool of a 15-month-old child with rotavirus diarrhoea and attenuated by tissue culture passaging. Other examples of diseases for which vaccines are produced from live attenuated microorganisms are the viral diseases measles, rubella and mumps, polio, yellow fever and influenza (an intranasal vaccine), and the bacterial diseases pertussis (whooping cough) and tuberculosis. In general, live attenuated vaccines are considered to be very immunogenic. To maintain their potency, they require special storage such as refrigerating and maintaining a cold chain. There is always a remote possibility that the attenuated bacteria or viruses mutate and become virulent (infectious).

In contrast, *inactivated vaccines* contain microorganisms whose DNA or RNA was first inactivated, so that they are 'dead' and cannot replicate and cause an infection anymore. Therefore, these vaccines are also safe in immune-compromised patients. Inactivation is usually achieved with heat or chemicals, such as formaldehyde or formalin, or radiation. There are several types of inactivated vaccines.

Whole inactivated vaccines are composed of entire viruses or bacteria. They are generally quite immunogenic. However, they are often also quite reactogenic (producing adverse events), which means that vaccinees may frequently suffer from local vaccine reactions at the site of vaccination (redness, itching, pain) or even from systemic vaccine reactions such as headache and fever. Fortunately, these reactions are usually benign, mild and transitory, and only last from hours to a few days. Whole vaccines have been developed for prophylaxis of, amongst others, pertussis (bacterial), cholera (bacterial) and influenza (viral).

Component vaccines do not contain whole microorganisms but preferably only those parts which have proven to stimulate the immune response most. The advantage of this approach is that other parts of the microorganism in question, which do not contribute to a relevant immune response but may cause unwanted vaccine reactions, can be removed (vaccine purification). Thus, component vaccines are usually less reactogenic than whole vaccines. Simple component vaccines are the split vaccines, which result after the treatment with membrane-dissolving liquids like such as ether. More sophisticated, subunit vaccines are produced using biological or genetic techniques. They essentially consist of a limited number of defined molecules, which can be found on the surface of microorganisms. Their vaccine reactogenicity is thereby further decreased. A disadvantage can be that isolated antigens may not stimulate the immune system as well as whole microorganisms. To overcome this problem, *virus-like particles* (VLP) *vaccines* and *liposomal vaccines* have been developed. Virus-like particles are particles that spontaneously assemble from viral surface proteins in the absence of other viral components. They mimic the structure of authentic spherical virus particles and they are believed to be more readily recognized by the immune system. In liposomal vaccines, the immunogenic subunits are incorporated into small vesicles sized as viruses (0.1–0.2 μm) and made of amphiphilic chemical compounds such as phospholipids (main components of biological membranes). Examples of component vaccines are *Haemophilus influenzae* type b (Hib) vaccines, hepatitis A and B vaccines, pneumococcal vaccines, and, again, influenza vaccines. The current generation of HPV vaccines are virus-like particles vaccines.

Another approach to increase the immunogenicity of inactivated vaccines is the use of *adjuvants*. These are agents that, by different mechanisms, augment the immune response against antigens. A potent adjuvant which has been used for over 50 years is aluminium hydroxide. In recent years, several new adjuvants have been developed: MF59 (an oil-in-water emulsion), MPL (a chemically modified derivative of lipopolysaccharide) and CpG 7909 (a synthetic nucleotide). Adjuvanted vaccines tend to more enhanced reactogenicity, i.e. they lead to higher incidences of local and systemic reactions. Some known adjuvants are therefore not suitable for human use (possibly still for veterinary use), for example, Freund's complete adjuvant (heat-killed *Mycobacterium tuberculosis* emulsified in mineral oil). This adjuvant is very effective to enhance both humoral and cellular immunity, but has been found to produce skin ulceration, necrosis and muscle lesion when administered as intramuscular injection. Other potential safety concerns of adjuvanted vaccines are immune-mediated adverse events (for example, anaphylaxis or arthritis) or chemical toxicity.

The immune system of infants and young children has difficulties to recognize those bacteria which have outer coats that disguise antigens. A notorious example is the bacterium *Streptococcus pneumoniae*. *Conjugate vaccines* may overcome this problem. While an adjuvanted vaccine consists of a physical mixture of vaccine and adjuvant, in a conjugate vaccine the microbial antigens are chemically bound to certain proteins or toxins (the carrier proteins), with the effect that recognition by the juvenile immune system is increased. This technique is used for Hib and pneumococcal vaccines.

Certain bacteria produce exotoxins capable of causing disease. Diphtheria is a bacterial disease, first described by Hippocrates (ca. 460–377 B.C.). Epidemics of diphtheria swept Europe in the seventeenth century and the American colonies in the eighteenth century. The causative bacterium is *Corynebacterium diphtheriae*, which produces diphtheria toxin. This toxin can be deprived of its toxic properties by inactivation with heat or chemicals, but it still carries its immunogenic properties; it is then called a toxoid and can be used for diphtheria *toxoid vaccine*. Another example is the tetanus vaccine containing the toxoid of the bacterium *Clostridium tetani*. Diphtheria and tetanus toxoid vaccines are often given to infants in combination with a vaccine for pertussis. This combination is known as DTP vaccine.

DTP vaccine is an example of a *combination vaccine*, which intends to prevent a number of different diseases, or one disease caused by different strains or different serotypes of the same species, such as the seasonal influenza vaccines which currently contain antigens of three or four virus (sub)types: one or two B-strains, an A-H1N1 strain and an A-H3N2 strain. Pneumococcal vaccines are currently available as 7-valent to even 23-valent vaccines. In contrast, a monovalent vaccine is intended to prevent one specific disease only caused by one defined microorganism, for example the hepatitis B vaccine.

1.4.2 Routes of Administration

Licensed vaccines differ with respect to the route of administration. This is not only a question of comfort for the vaccinee but also depends on the exact types and location of immune cells to which the vaccine is offered to achieve the optimal prophylactic effect. *Injectable vaccines* are usually given subcutaneously (into the fat layer between skin and muscle) or intramuscularly (directly into a muscle). Preferred vaccination sites are the deltoid region of the arm in adults and elderly, and the thigh in newborns and infants. Some vaccines—hepatitis B vaccines, for example—can also be administered intramuscularly in the buttock. An alternative to subcutaneous/intramuscular injection is *intradermal vaccination*, directly into the dermis. Intradermal vaccination is successfully used for rabies and hepatitis B. In case of influenza, it reduces the dose needed to be given. This route could thus increase the number of available doses of vaccine, which can be relevant in case of an influenza pandemic. Vaccination by injection is often felt to be uncomfortable by vaccinees, it usually needs some formal medical training to administer it, and it carries the risk of needle prick accidents with contaminated blood.

An alternative is administration by the oral route, since the 1950s used for the live attenuated polio vaccine: some droplets of vaccine-containing liquid on a lump of sugar to be swallowed. This route builds up a strong local immunity in the intestines, the site of poliovirus entry. Obvious advantages are the increased ease and acceptance of vaccination and the absence of the risk of blood contamination.

The third option is intranasal vaccine administration, preferably used for respiratory pathogens. *Intranasal vaccines* are dropped or sprayed into the cavity of the nose. Advantages are, again, the ease of administration (in particular for childhood vaccines), the direct reach of the respiratory compartment, and hence induction of local protective immunity at the primary site of pathogen entry.

1.4.3 Leaky Vaccines Versus All-or-Nothing Vaccines

Leaky vaccines (also partial vaccines) are vaccines that reduce but do not eliminate a vaccinated person's risk of infection upon exposure to the pathogen, meaning that the protection against infection is incomplete. In contrast, *all-or-nothing vaccines* reduce infection rates to zero for some fraction of subjects while the remaining fraction is fully susceptible. The subjects who after vaccination are no longer at risk for the disease are said to be *completely protected.* Leaky vaccines protect subjects with fewer exposures at a higher rate than subjects with more exposures. Examples of leaky vaccines are influenza vaccines and pertussis vaccines. Examples of vaccines that are assumed to function as all-or-nothing vaccines are measles vaccines and rubella vaccines.

1.4.4 Malaria Vaccines

Malaria is an example of a protozoan disease. The most serious forms of the disease are caused by the parasites *Plasmodium falciparum* and *Plasmodium vivax*. The parasites are transmitted by the female Anopheles mosquito. Sporozoites (from *sporos*, seed) of the parasites are injected in the bitten person. In the liver, sporozoites develop into blood-stage parasites which then reach red blood cells. There are three types of malaria vaccines in development: pre-erythrocytic vaccines, blood-stage vaccines and transmission-blocking vaccines. Pre-erythrocytic vaccines target the sporozoites and the liver life forms. If fully effective, they would prevent blood-stage infection. In practice, they will be only partially effective, but they may reduce the *parasite density* (density of malaria parasites in the peripheral blood) in the initial blood-stage of the disease. Blood-stage vaccines try to inhibit parasite replication by binding to the antigens on the surface of infected red blood cells. These vaccines also may reduce parasite density to a level that prevents development of clinical disease. Transmission-blocking vaccines try to prevent transmission of the parasite to humans rather than preventing infection. This is attempted by trying to induce antibodies that act against the sexual stages of the parasite, to prevent it from becoming sexually mature.

1.4.5 Experimental Prophylactic and Therapeutic Vaccines

Some recent developments in vaccine research, still in an experimental stage in animal models, are recombinant vector vaccines and DNA vaccines. *Recombinant vector vaccines* are vaccines created by recombinant DNA technology. The pathogen's DNA is inserted into a suitable virus or bacterium that transports the DNA into healthy body cells where the foreign DNA is read. Consequently, foreign proteins are synthesized and released, which act as antigens stimulating an immune response. Similarly, *DNA vaccines* are made of plasmids, circular pieces of bacterial DNA with incorporated genetic information to produce an antigen of a pathogen. When the vaccine DNA is brought into suitable body cells, the antigen is expressed, and the immune system can respond to it. The advantage of DNA vaccines is that no outer source of protein antigen is needed. Serious safety concerns will have to be addressed before these experimental approaches can be tested in man.

The vaccines discussed so far were all *prophylactic vaccines*, intended to prevent infection. A fairly recent development is the emergence of *therapeutic vaccines*, not given with the intention to prevent but to treat. The targeted diseases need not to be infectious. Therapeutic tumour vaccines, for example, are aimed at tumour forms that the immune system cannot destroy. The hope is to stimulate the immune system in such a way that the enhanced immune response is able to kill the tumour cells. Therapeutic tumour vaccines are being developed for acute myelogenous leukaemia, breast cancer, chronic myeloid leukaemia, colorectal cancer, oesophageal cancer, head and neck cancer, liver and lung cancers, melanoma, non-Hodgkin lymphoma and ovarian, pancreatic and prostate cancers. Other examples of therapeutic vaccines being developed are addiction vaccines for cocaine and nicotine abuse. Nicotine is made of small molecules that are able to pass the blood–brain barrier, a filter to protect the brain from dangerous substances. One vaccine in development has the effect that the subject develops antibodies to nicotine, so that when they smoke, the antibodies attach to the nicotine and make the resulting molecule too big to pass the blood–brain barrier, so that smoking stops being pleasurable. Another nicotine vaccine in development leads to the production of antibodies that block the receptor that is involved in smoking addiction. Therapeutic vaccines are also being tested for hyperlipidaemia, hypertension, multiple sclerosis, rheumatoid arthritis and Parkinson's disease.

Chapter 2
Humoral and Cellular Immunity

Abstract This chapter offers a clear account of humoral immunity, the component of the immune system involving antibodies that circulate in the humor, and cellular immunity, the component that provides immunity by action of cells. Antibody titres and antibody concentrations are looked at, and two standard assays for humoral immunity, the haemagglutination inhibition test and ELISA, are introduced. Standard assays for cellular immunity are briefly looked at including the ELISPOT assay.

2.1 Humoral Immunity

When the adaptive immune system is activated by the innate immune system, the *humoral immune response* (also *antibody-mediated immune response*) triggers specific B cells to develop into plasma cells. These plasma cells then secrete large amounts of antibodies. Antibodies circulate in the lymph and the bloodstreams. (Hence the name, humoral immunity. *Humoral* comes from the Greek *chymos*, a key concept in ancient Greek medicine. In this view, people were made out of four fluids: blood, black bile, yellow bile and mucus (phlegm). Being healthy meant that the four humors were balanced. Having too much of a humor meant unbalance resulting in illness.) The more general term for antibody is *immunoglobulin*, a group of proteins. There are five different antibody classes: IgG, IgM, IgA, IgE and IgD. The first three, IgG, IgM and IgA, are involved in defence against viruses, bacteria and toxins. IgE is involved in allergies and defence against parasites. IgD has no apparent role in defence. The primary humoral immune response is usually weak and transient, and has a major IgM component. The secondary humoral response is stronger and more sustained and has a major IgG component.

Antibodies attack the invading pathogens. Different antibodies can have different functions. One function is to bind to the antigens and mark the pathogens for destruction by phagocytes, which are cells that phagocytose (ingest) harmful microorganism and dead or dying cells. Some antibodies, when bound to antigens, activate the *complement*, serum proteins able to destroy pathogens or to induce the destruction of pathogens. These antibodies are called *complement-mediated antibodies*. *Neutralizing antibodies* are antibodies that bind to antigens so that the antigen can no longer

J. Nauta, *Statistics in Clinical and Observational Vaccine Studies*,
Springer Series in Pharmaceutical Statistics,
https://doi.org/10.1007/978-3-030-37693-2_2

recognize host cells, and infection of the cells is inhibited. For example, in case of a virus, neutralizing antibodies bind to viral antigens and prevent the virus from attachment to host cell receptors.

It is good practice to state the antigen against which the antibody was produced: anti-HA antibody, anti-tetanus antibody, anti-HPV antibody, etc.

2.1.1 *Antibody Titres and Antibody Concentrations*

Antibody levels in serum samples are measured either as antibody titres or as antibody concentrations. An *antibody titre* is a measure of the antibody amount in a serum sample, expressed as the reciprocal of the highest dilution of the sample that still gives (or still does not give) a certain assay read-out. To determine the antibody titre, a serum sample is *serially* (stepwise) *diluted.* The *dilution factor* is the final volume divided by the initial volume of the solution being diluted. Usually, the dilution factor at each step is constant. Often used dilution factors are 2, 5 and 10. In this book, the *starting dilution* will be denoted by $1{:}D$. A starting dilution of 1:8 and a dilution factor of 2 will result in the following twofold serial dilutions: 1:8, 1:16, 1:32, 1:64, 1:128, and so on. To each dilution, a standard amount of antigen is added. An assay (test) is performed which gives a specified read-out either when antibodies against the antigen are detected or, depending on the test, when no antibodies are detected. The higher the amount of antibody in the serum sample, the higher the dilutions at which the assay read-out occurs (or no longer occurs). Assume that the assay read-out occurs for the dilutions 1:8, 1:16 and 1:32, but not for the dilutions 1:64, 1:128, etc. The antibody titre is the reciprocal of the highest dilution at which the read-out did occur, 32 in the example. If the assay read-out does not occur at the starting dilution—indicating a very low number of antibodies, below the detection limit of the assay—then often the antibody titre for the sample is set to $D/2$, half of the starting dilution. By definition, antibody titres are dimensionless.

Antibody concentrations measure the amount of antibody-specific protein per millilitre serum, expressed either as micrograms of protein per millilitre (μg/ml) or as units per millilitre (U/ml). (A unit is an arbitrary amount of a substance agreed upon by scientists.) The measurement of antibody concentrations is usually done on a single serum sample rather than on a range of serum dilutions.

2.1.2 *Two Assays for Humoral Immunity*

To give the reader an idea of how antibody levels in serum samples are determined, below two standard assays for humoral immunity are discussed, the haemagglutination inhibition test involving serum dilutions, and the enzyme-linked immunosorbent assay involving a single serum.

Some viruses—influenza, measles and rubella, amongst others—carry on their surface a protein called *haemagglutinin* (HA). When mixed with erythrocytes (red blood cells) in an appropriate ratio, it causes the blood cells clump together (agglutinate). This is called *haemagglutination*. Anti-HA antibodies can inhibit (prevent) this reaction. This effect is the basis for the *haemagglutination inhibition* (HI, also HAI) assay, an assay to determine antibody titres against viral haemagglutinin. First, serial dilutions of the antibody-containing serum are allowed to react with a constant amount of antigen (virus). In the starting dilution and the lower dilutions, the amount of antibody is larger than the amount of antigen, which means that all virus particles are bound by antibody. At a certain dilution, the antibody amount becomes smaller than the antigen amount, which means that free, unbound virus remains. This free antigen is then detected by the second part of the test: to all dilutions, a defined amount of erythrocytes is added. In the lower dilutions, where all antigen is bound by antibody, the erythrocytes freely sink to the lowest point of the test tube or well and form a red spot there (no haemagglutination). In higher dilutions, where there is so less antibody that free virus remains, this virus binds to erythrocytes, which then form a wide layer in the test tube (haemagglutination). The reciprocal of the last dilution where haemagglutination is still inhibited (i.e. where haemagglutination does not occur) is the antibody titre.

The *enzyme-linked immunosorbent assay* (ELISA), also called *enzyme immunoassay* (EIA), is another assay to detect the presence of antibodies in a serum sample. Many variants of the test exist, and here only the basic principle will be explained. In simple terms, a defined amount of antigen is bound to a solid-phase surface, usually the plastic of the wells of a microtitre plate. Then a serum sample with an unknown amount of antigen-specific antibody is added and allowed to react. If antibody is present, it will bind to the fixed antigen. Consequently, the serum (with unbound antibody, if any) is washed away, while the fixed antigen–antibody complexes remain on the solid-phase surface. They are detected by adding a solution of antibodies against human immunoglobulin, prepared in animals and chemically linked to an enzyme. The fixed complexes consist of three components: the test antigen, the antibody of unknown amount from the serum specimen and the enzyme-labelled secondary test antibody against the serum antibody. A substrate to the enzyme is added, which is split by the enzyme, if present. One of the released splitting products can give a detectable signal, a certain colour, for example. Only if the three-component complex is present (i.e. if there has been antibody in the serum specimen), this signal will occur. The strength of the signal is a measure of the amount of serum antibody.

The first-generation ELISA uses chromogenic substrates, which release colour molecules after enzymatic reaction. By a spectrophotometer, the intensity of the colour in the solution (or the amount of light absorbed by the solution) can be determined (optical density). The antibody concentration is determined by comparing the optical density of the serum sample with an optical density curve constructed with the help of a standard sample.

In a fluorescence ELISA, which has a higher sensitivity than a colour-releasing ELISA, the signal is given by fluorescent molecules, whose amount can be measured by a spectrofluorometer.

2.2 Cellular Immunity

Cellular immunity (also *cell-mediated immunity* (CMI)) is an adaptive immune response that is primarily mediated by thymus-derived small lymphocytes, which are known as T cells. Here, two types of T cells are considered: T helper cells and T killer cells. *T helper cells* are particularly important because they maximize the capabilities of the immune system. They do not destroy infected cells or pathogens, but they activate and direct other immune cells to do so, which explains their name: T helper cells. The major roles of T helper cells are to stimulate B cells to secrete antibodies, to activate phagocytes, to activate T killer cells and to enhance the activity of natural killer (NK) cells. Another term for T helper cells is CD4+ T cells (CD4 positive T cells), because they express the surface protein CD4. T helper cells are subdivided on the basis of the cytokines they secrete after encountering a pathogen. *T Helper 1 cells* (TH1 cells) secrete many different types of cytokines, the principal being interferon-γ (IFN-γ), interleukin-2 (IL-2) and interleukin-12 (IL-12). IFN-γ has many effects including activation of macrophages to deal with intracellular bacteria and parasites. IL-2 stimulates the maturation of killer T cells and enhances the cytotoxicity of NK cells. IL-12 induces the secretion of INF-γ. The principal cytokines secreted by *T helper 2 cells* (TH2 cells) are interleukin-4 (IL-4) and interleukin-5 (IL-5) for helping B cells. An infection with the human immunodeficiency virus (HIV) demonstrates the importance of helper T cells. The virus infects CD4+ T cells. During an HIV infection, the number of CD4+ T cells drops, leading to the disease known as the acquired immune deficiency syndrome (AIDS).

The major function of *T killer cells* is cytotoxicity to recognize and destroy cells infected by viruses, but they also play a role in the defence against intracellular bacteria and certain types of cancers. Intracellular pathogens are usually not detected by macrophages and antibodies, and clearance of infection depends upon elimination of infected cells by cytotoxic lymphocytes. T killer cells are specific, in the sense that they recognize specific antigens. Alternative terms for T killer cells are CD8+ T cells (CD8 positive T cells), cytotoxic T cells and CTLs (cytotoxic T lymphocytes). CD8+ T cells secrete INF-γ and the inflammatory cytokine tumour necrosis factor (TNF).

Most assays for cellular immunity are based on cytokine secretion, as marker of T cell response. A wide variety of assays exists, but the most used one is the enzyme-linked immunospot (ELISPOT) assay, which was originally developed as a method to determine the number of B cells secreting antibodies. Later the method was adapted to determine the number of T cells secreting cytokines. ELISPOT assays are performed in microtitre plates coated with the relevant antigen. *Peripheral blood mononuclear cells* (PBMCs) are added to it and then incubated. (PBMCs are white blood cells such as lymphocytes and monocytes.) When the cells are secreting the specific cytokine, discrete coloured spots are formed, which can be counted. One of the most popular of this type of assays to evaluate cellular immune responses is the INF-γ ELISPOT assay, an assay for CTL activity. Results are expressed as *spot-forming cells* (SPCs) per million peripheral blood mononuclear cells (SPC/10^6 PMBC). Other types of the assay are the IL-2 ELISPOT assay, the IL-4 ELISPOT assay, etc.

The *FluoroSpot assay* is a modification of the ELISPOT assay and is based on using multiple fluorescent anticytokines, which makes it possible to spot two cytokines in the same assay.

Other assays that can quantitate the number of antigen-specific T cells are the intracellular cytokine assay and the tetramer assay.

Flow cytometry uses the principles of light scattering and emission of fluorochrome molecules to count cells. Cells are labelled with a fluorochrome, a fluorescent dye used to stain biological specimens. A solution with cells is injected into the flow cytometer, and the cells are then forced into a stream of single cells by means of hydrodynamic focusing. When the cells intercept light from a source, usually a laser, they scatter light and fluorochromes are realized. Energy is released as a photon of light with specific spectral properties unique to the fluorochrome. By using different colours of fluorescent labelling, a single assay can quantify different types of cells.

Part II
Analysis of Immunogenicity Data

Chapter 3
Standard Statistical Methods for Immunogenicity Data

Abstract In this chapter, the four standard statistics to summarize humoral and cellular immunogenicity data—the geometric mean response, the geometric mean fold increase, the seroprotection rate and the seroconversion rate—are introduced. For each of these summary statistics, the standard statistical analysis is covered in great detail, both for single vaccine and comparative vaccine trials. A persistent misconception about the geometric mean fold increase and baseline imbalance is ironed out. The analysis of rates (i.e. proportions) receives extensive treatment, because binary immunogenicity and safety endpoints are very common in clinical vaccine research. A simple but very effective bias correction for the rate ratio is brought to the attention of the reader. Superior performing alternatives to the conventional methods for comparing two rates are given a prominent place. It is explained how exact confidence intervals for a risk difference or a relative risk can be obtained. The reverse cumulative distribution plot, an intuitively pleasing and useful graphic tool to display immune response profiles, is exemplified. Statistical tests that can be applied to compare reverse cumulative distribution curves are presented.

3.1 Introduction

There is an ancient proverb, popularized by Spanish novelist Cervantes in his *Don Quixote* (1605), that says that the proof of the pudding is in the eating. Putting it figuratively, ideas and theories should be judged by testing them. For vaccines, the test is the vaccine efficacy study. A group of infection-free subjects are randomized to be vaccinated with either the investigational vaccine or a placebo vaccine. The subjects are then followed-up, to monitor how many cases of the infectious disease occur in each of the two arms of the trial. If in the investigational arm the number of cases is significantly lower than in the placebo arm, this is considered to be proof that the investigational vaccine protects from infection. Vaccine efficacy studies, however, have a notorious reputation among vaccine researchers. They are extremely if not prohibitively costly, as they usually require large sample sizes and a lengthy follow-up. If during the follow-up period the attack rate of the infection and thus the number of cases is low, the period has to be extended, meaning even higher costs. Many

J. Nauta, *Statistics in Clinical and Observational Vaccine Studies*, Springer Series in Pharmaceutical Statistics,
https://doi.org/10.1007/978-3-030-37693-2_3

vaccine efficacy studies have been negative as a result of imperfect case finding. Placebo-controlled vaccine efficacy studies in elderly are considered unethical if for the infectious disease a vaccine is available.

A popular alternative to vaccine efficacy studies are *vaccine immunogenicity trials*. In such trials, the primary endpoint is a humoral or a cellular immunity measurement which is known to correlate with protection against infection (Chap. 11.) Vaccine immunogenicity trials are usually much smaller and require often only a short follow-up, which makes them less costly than vaccine efficacy studies. Registration authorities such as the United States Food and Drug Administration (FDA), the European Medicines Agency (EMA), Japan's Organization for Pharmaceutical Safety and Research (OPSR), China's National Medical Products Administration (NMPA) and the National Registration Authority (NRA) of Australia all accept the results of vaccine immunogenicity trials in support of licensure of new vaccines. In early vaccine development, they may be used to explore dose formulation. They may be used to expand the use of a vaccine by extending the age eligible for the vaccine. This was done to extend the lower age range of an HPV vaccine down to age 9 from 11. Also, when a vaccine is updated, by adding serotypes to an existing vaccine, immunogenicity studies can play an important role. In paediatric studies where children are expected to take a several vaccines, immunogenicity assays are often used to check that the new vaccine doesn't impact immunogenicity of already recommended vaccines. Finally, in assessing the long-term effect of a vaccine, some kind of immunogenicity assay is used.

Until recently vaccine immunogenicity trials typically focused on the humoral immune response, i.e. on serum antibody levels. Today, many papers on vaccine immunogenicity report also on cellular immunity. Nevertheless, cellular immunity in vaccine trials is still largely in the investigational phase.

There are four standard statistics to summarize humoral and cellular immunity data: the geometric mean response, the geometric mean fold increase, the seroprotection rate and the seroconversion rate. Two of these statistics, the geometric mean response and the seroprotection rate, quantify absolute immunogenicity values, while the other two, the geometric mean fold increase and the seroconversion rate, quantify intra-individual changes in values. In the next sections, the statistical analysis of these four summary statistics is discussed, both for single vaccine groups and for two vaccine groups.

3.2 Geometric Mean Titres and Concentrations

Distributions of post-vaccination humoral and cellular immunogenicity values tend to be skewed to the right. Log-transformed immunogenicity values, on the other hand, usually are approximately normally distributed. Thus, standard statistical methods requiring normal data can be applied to the log-transformed values. Antilogs of point and interval estimates can be used for inference about parameters of the distribution underlying the untransformed values.

The standard statistics to summarize immunogenicity values is the *geometric mean titre* (GMT) if the observations are titres, or the *geometric mean concentration* (GMC) if the observations are concentrations. Let $v_1, \ldots, v_n$ be a group of n immunogenicity values. (Throughout this book, groups of observations are assumed to be independent and identically distributed (i.i.d.).) The geometric mean is defined as

$$GM = (v_1 \times \cdots \times v_n)^{1/n}.$$

An equivalent expression is

$$GM = \exp\left(\sum_{i=1}^{n} \log v_i/n\right).$$

The geometric mean response is thus on the same scale as the immunogenicity measurements.

The transformation of the immunogenicity values need not to be $\log_e$, it can be any logarithmic transformation, $\log_2$, $\log_{10}$, etc. Care should be taken that when calculating the geometric mean response the correct base is used. Thus, if $\log_{10}$ is used, the geometric mean should be calculated as

$$GM = 10^{\sum_{i=1}^{n} \log_{10} v_i/n}.$$

If antibody titres t_i are reciprocals of twofold serial dilutions with 1:D as the lowest tested dilution, then a convenient log transformation is

$$u_i = \log_2[t_i/(\mathrm{D}/2)]. \tag{3.1}$$

The u_i's are then the dilution steps: 1, 2, 3, etc. The geometric mean should be calculated as

$$GM = (\mathrm{D}/2)2^{\sum_{i=1}^{n} u_i/n}.$$

Transformation (3.1) will be referred to as the *standard log transformation for antibody titres*.

Example 3.1 Rubella (German measles) is a disease caused by the rubella virus. In adults, the disease itself is not serious, but infection of a pregnant woman by rubella can cause miscarriage, stillbirth or damage to the foetus during the first three months of pregnancy. A haemagglutination inhibition (HI) test for rubella is often performed routinely on pregnant women. The presence of a detectable HI titre indicates previous infection and immunity to reinfection. If no antibodies can be detected, the woman is considered susceptible and is followed accordingly. Assume that in the HI test the lowest dilution is 1:8. Then the HI titres can take on the values 8, 16, 32, etc. The standard log-transformed values are $\log_2(8/4) = 1$, $\log_2(16/4) = 2$, $\log_2(32/4) = 3$, etc. The geometric mean of the five titres 8, 8, 16, 32, 64 is

$$GMT = 4 \times 2^{(1+1+2+3+4)/5} = 18.379.$$

With the standard log transformation, differences between log-transformed values are easy to interpret: a difference of 1 means a difference of one dilution, a difference of 2 means a difference of two dilutions, etc.

A statistic often reported with the geometric mean response is the *geometric standard deviation* (GSD), which is the antilog of the sample standard deviation of the log transformed immunogenicity values. The statistic allows easy calculation of confidence limits for the geometric mean of the distribution underlying the immunogenicity values (the underlying geometric mean for short). Let *SD* be the sample standard deviation of the log transformed immunogenicity values, then the geometric standard deviation is

$$GSD = \exp(SD).$$

The lower and upper limit of the two-sided 100(1−α)% confidence interval for the underlying geometric mean e^{μ} are

$$LCL_{e^{\mu}} = GMT / GSD^{t_{n-1;1-\alpha/2}/\sqrt{n}} \tag{3.2}$$

and

$$UCL_{e^{\mu}} = GMT \times GSD^{t_{n-1;1-\alpha/2}/\sqrt{n}}, \tag{3.3}$$

where $t_{n-1;1-\alpha/2}$ is the 100(1−α/2)th percentile of Student's t-distribution with $(n-1)$ degrees of freedom.

Example 3.1 (continued) The sample standard deviation *SD* of the five log-transformed HI titres is 0.904. Thus, the geometric standard deviation is

$$GSD = \exp(0.904) = 2.469.$$

Percentiles of Student's t-distribution can be obtained with the SAS function TINV. The lower 95% confidence limit for the underlying geometric mean is

$$LCL_{e^{\mu}} = 18.379/2.469^{2.776/\sqrt{5}} = 5.98$$

and the upper 95% confidence limit is

$$UCL_{e^{\mu}} = 18.379 \times 2.469^{2.776/\sqrt{5}} = 56.4,$$

where 2.776 = TINV(0.975,4).

3.2.1 Single Vaccine Group

If the $u_i = \log v_i$ are normally distributed with mean μ and variance σ^2, then the arithmetic mean of the u_i is a point estimate of μ, and

$$GMT = \exp\left(\sum_{i=1}^{n} u_i/n\right)$$

a point estimate of e^{μ}, the underlying geometric mean. The distribution of the u_i is known as the log-normal distribution.

Because the u_i are normally distributed, confidence intervals for μ can be based on the one-sample t-test. Antilogs of the limits of the t-test based $100(1-\alpha)\%$ confidence interval for μ constitute $100(1-\alpha)\%$ confidence limits for the parameter e^{μ}. These confidence limits are identical to those in (3.2) and (3.3). A nice property of the log-normal distribution is that e^{μ} is not only its geometric mean but also its median.

Example 3.2 The following data are six Th1-type IFN-γ values: 3.51, 9.24, 13.7, 35.2, 47.4 and 57.5 IU/L. The natural logarithms are 1.256, 2.224, 2.617, 3.561, 3.859 and 4.052, with arithmetic mean 2.928 and standard error 0.444. Hence,

$$GMC = \exp(2.928) = 18.7.$$

With $t_{0.975,5} = 2.571$, it follows that the two-sided 95% confidence limits for μ are

$$2.928 - 2.571(0.444) = 1.786$$

and

$$2.928 + 2.571(0.444) = 4.070.$$

Thus, the lower and upper 95% confidence limits for the geometric mean e^{μ} of the distribution underlying the IFN-γ values are $e^{1.786} = 5.97$ and $e^{4.070} = 58.6$.

By definition, confidence intervals for geometric mean responses are non-symmetrical.

3.2.2 Two Vaccine Groups

If there are two vaccine groups, statistical inference is based on the two-sample t-test, applied to the log-transformed immunogenicity values. Point and interval estimates for the difference $\Delta = \mu_1 - \mu_0$ are transformed back to point and interval estimates for the ratio $\theta = e^{\mu_1}/e^{\mu_0}$.

The standard statistic to compare two groups of immunogenicity values is the *geometric mean ratio* (GMR):

$$GMR = GM_1/GM_0,$$

where GM_1 and GM_0 are the geometric mean response of investigational and the control vaccine group, respectively. Let AM_1 and AM_0 denote the arithmetic means of the log-transformed values of the two groups, then the following equality holds:

$$GMR = \exp(AM_1 - AM_0).$$

Thus, the P-value from the two-sample t-test to test the null hypothesis $H_0 : \Delta = 0$ can be used to test the null hypothesis $H_0 : \theta = 1$.

Example 3.2 (continued) Assume that the six Th1-type IFN-γ values are to be compared with a second group of six values, and that the arithmetic mean and standard error of the log-transformed values are 2.754 and 0.512, respectively. Thus,

$$GMC_0 = \exp(2.754) = 15.7,$$

and the geometric mean ratio is

$$\begin{aligned} GMR &= \exp(2.928 - 2.754) = \exp(0.174) \\ &= 1.19 = 18.7/15.7. \end{aligned}$$

The estimated standard error of the difference is 0.678. Lower and upper 95% confidence limits for the underlying geometric mean ratio are obtained as

$$\exp[1.19 - 2.228(0.677)] = 0.727$$

and

$$\exp[1.19 + 2.228(0.677)] = 14.855,$$

where 2.228 = TINV(0.975,10).

3.3 Geometric Mean Fold Increase

For some infectious diseases, pre-vaccination immunogenicity levels are not zero. An example is influenza. Recipients of influenza vaccines have usually been exposed to various influenza viruses during lifetime, by natural infections or previous vaccinations (exceptions are very young children). In that case, post-vaccination immunogenicity levels do not only express the immune responses to the vaccination but also the pre-vaccination levels. In that case, an alternative to the geometric mean titre or concentration is the geometric mean fold increase (also mean fold increase, geometric mean fold rise).

If v_{pre} is a subject's pre-vaccination (baseline) immunogenicity value and v_{post} the post-vaccination value, then the *fold increase* is

$$f = v_{post}/v_{pre}.$$

Fold increases express intra-individual relative increases in immunogenicity values. Just like immunogenicity values, log-transformed fold increases tend to be normally distributed, and for the statistical analysis of fold increases, the methods described above for the analysis of immunogenicity values can be used.

3.3.1 Analysis of a Single Geometric Mean Fold Increase

The standard statistic to summarize a group of n fold increases $f_1, \ldots, f_n$ is the *geometric mean fold increase* (gMFI)

$$gMFI = \exp\left(\sum_{i=1}^{n} \log f_i/n\right).$$

It is easy to show that

$$gMFI = GM_{post}/GM_{pre}.$$

Thus, the geometric mean fold increase is identical to the geometric mean of the post-vaccination values divided by the geometric mean of the pre-vaccination values. Note, though, that this equation only holds if for all n subjects both the pre- and the post-vaccination value is non-missing.

Example 3.3 Consider an influenza trial in which pre- and post-vaccination anti-HA antibody levels are measured by means of the HI test. Let (5,40), (5,80), (10,160), (10,320), (20,80) and (20,640) be the pre- and post-vaccination antibody titres of the first six subjects enrolled. $GMT_{pre} = 10.0$ and $GMT_{post} = 142.5$. The fold increases are 8, 16, 16, 32, 4 and 32. The geometric mean of these sixfold increases is $gMFI = 14.25$. The same value is obtained if GMT_{post} is divided by GMT_{pre}. The geometric standard deviation of the fold increases is $GSD = 2.249$. Forms. (3.2) and (3.3) can be used to obtain a confidence interval for the geometric mean of the distribution underlying the fold increases.

3.3.2 Analysis of Two Geometric Mean Fold Increases

To compare two groups of fold increases, the two-sample t-test can be applied to the log-transformed fold increases.

In case of two groups of fold increases, the following equation holds

$$\frac{gMFI_1}{gMFI_0} = \frac{GM_{\text{post 1}}/GM_{\text{pre 1}}}{GM_{\text{post 0}}/GM_{\text{pre 0}}} = \frac{GM_{\text{post 1}}/GM_{\text{post 0}}}{GM_{\text{pre 1}}/GM_{\text{pre 0}}} = \frac{GMR_{\text{post}}}{GMR_{\text{pre}}}.$$

Thus, the ratio of two mean fold increases (the *geometric mean fold ratio* (gMFR)) is identical to geometric mean ratio of the post-vaccination immunogenicity values divided by the geometric mean ratio of the pre-vaccination immunogenicity values. This observation has an interesting implication. Consider a randomized trial in which two vaccines are being compared. Because of the randomization, and if the sample sizes are not too small, $GM_{\text{pre 0}}$ will approximately be equal to $GM_{\text{pre 1}}$, and their ratio GMR_{pre} will be approximately equal to 1.0. Thus, the ratio of the geometric mean fold increases will be approximately equal to the ratio of the geometric means of the post-vaccination immunogenicity values:

$$gMFR = \frac{gMFI_1}{gMFI_0} \approx GMR_{\text{post}}.$$

In other words, if there is no baseline imbalance, an analysis of the fold increases will yield a result virtually identical to that of the analysis of the post-vaccination immunogenicity values.

Example 3.3 (continued) Assume that in the trial two influenza vaccines are being compared. Let (5,40), (5,80), (10,80), (10,80) and (5,80), (5,80), (10,80), (10,160) be the pre- and post-vaccination antibody titres of the experimental and the control vaccine group, respectively. The following summary statistics are found

$$GMT_{\text{pre 1}} = GMT_{\text{pre 0}} = 7.1$$

and

$$GMT_{\text{post 1}} = 67.27 \quad \text{and} \quad GMT_{\text{post 0}} = 95.14.$$

The fold increases are: 8, 16, 8, 8 and 16, 16, 8, 16. The mean fold increases are

$$gMFI_1 = 9.51 \quad \text{and} \quad gMFI_0 = 13.45.$$

Thus, the ratio of the mean fold increases

$$gMFR = 13.45/9.51 = 1.4,$$

which is indeed identical to the ratio of the post-vaccination geometric mean titres

$$GMR_{\text{post}} = 95.14/67.27 = 1.4.$$

As explained, the reason why these two ratios are equal is that the two pre-vaccination geometric mean titres are approximately equal.

If there is no baseline imbalance between two vaccine groups, it will be inefficient to use the fold increases to analyse post-vaccination immunogenicity levels, because, in general, fold increases are more variable than post-vaccination immunogenicity values, with as result in larger P-values and wider confidence intervals. The explanation for this is that the log-transformed fold increase is a difference, namely, between a log-transformed post-vaccination and a log-transformed pre-vaccination immunogenicity value. If the variances σ^2 at baseline and post-vaccination are the same, then the variance of the log-difference is $2\sigma^2(1-\rho)$, which implies that if the correlation ρ between the post- and the pre-vaccination values is less than 0.5, the variance of the difference is larger than that of the post-vaccination values.

3.3.3 A Misconception About Fold Increases and Baseline Imbalance

A change score is an intra-individual difference between a post- and a pre-treatment value. On a logarithmic scale, a fold increase is a change score:

$$\begin{aligned}\log f &= \log(v_{\text{post}}/v_{\text{pre}}) \\ &= \log v_{\text{post}} - \log v_{\text{pre}}.\end{aligned}$$

If in a clinical trial the baseline values of a given characteristic (age or weight, for example) or a measurement (say, a bioassay) differ between the treatment groups, then it is said that there is baseline imbalance. Baseline imbalance matters only if the baseline value measurement is related to the primary endpoint, i.e. if it is prognostic. In that case, one treatment will have a poorer prognosis than the other. The effect of the imbalance depends on the size of the imbalance and the strength of the association between the baseline value and the endpoint. If the characteristic or the baseline measurement is not prognostic, then baseline imbalance is of no concern and can be ignored.

For many bioassays, non-zero (detectable) pre-vaccination values will be predictive of the post-vaccination values because pre- and post-vaccination bioassay values tend to be positively correlated. It is often argued that imbalance in baseline values of a measurement can be dealt with by change scores, like fold increases, for example. The reasoning being that if the pre-treatment values are subtracted from the post-treatment values, any bias due to baseline imbalance is eliminated. This is a fallacy. Change scores are generally correlated with pre-treatment values, the correlation often being negative. For bioassays this phenomenon is also often seen, the higher the average pre-vaccination value the smaller the average fold increase. A difference between vaccine groups in pre-vaccination state is thus predictive not only of a difference in post-vaccination state but also of a difference in fold increase, albeit in the opposite direction. Hence, in case of baseline imbalance, a comparison of fold increases is in favour of the vaccine group with the smaller pre-vaccination values. Thus, it is a misconception that a fold increase analysis deals with baseline imbalance. A proper statistical technique to control for pre-vaccination state is analysis of covariance, which is discussed in Chap. 5 of this book.

3.4 Two Seroresponse Rates

3.4.1 Seroprotection Rate

For many infectious diseases, it is assumed that there is a given antibody level that is associated with protection from infection or disease. This antibody level is called the *threshold of protection*, and a subject is said to be *seroprotected* if the antibody level is equal to or above the threshold. For influenza, for example, the threshold is an anti-HA antibody level of 40, and subjects with an anti-HA antibody titre ≥40 are seroprotected for influenza. For diphtheria and tetanus, a protection threshold found in the literature is an anti-D/anti-T antibody concentration of 0.1 IU/ml. Seroprotection is thus a binary endpoint, and the *seroprotection rate* is the proportion of vaccinated subjects who are seroprotected.

What is meant with 'associated with protection' is not always clearly defined or understood. Sometimes it is interpreted as meaning that being seroprotected implies being fully protected against the disease. It is this, incorrect, interpretation that may have led to an overestimation of the importance of the concept of seroprotection. Being seroprotected means a moderate to high probability of protection. For example, there is evidence that subjects with an anti-HA antibody titre of 40 have a probability of 0.5 of being protected against influenza, when exposed to the virus [2].

3.4.2 Seroconversion Rate

Stedman's Medical Dictionary defines *seroconversion* as development of detectable particular antibodies in the serum as a result of infection or immunization. A subject without antibodies in his serum is called *seronegative*, while a subject with antibodies is called *seropositive*. A subject's *serostatus* is his status with respect to being seropositive or seronegative for a particular antibody. A subject whose serostatus was seronegative but after immunization became seropositive is said to have *seroconverted*. In a report on the safety and immunogenicity of a live attenuated human rotavirus vaccine, Vesikari et al. define seroconversion as appearance of serum IgA to rotavirus in post-vaccination sera at a titre of ≥20 U/ml in previously uninfected infants [3]. Depending on the vaccine dose, 73–96% of the infant subjects seroconverted.

In the scientific literature, however, alternative definitions of seroconversion can be found. A popular alternative definition is: a fourfold rise (also increase) in antibody level. This definition is often used when recipients of a vaccine may be seropositive at enrolment. The major cause of cervical cancer and cervical dysplasia (abnormal maturation of cells within tissue) is the human papillomavirus (HPV). Cervical cancer is cancer of the cervix, the lower part of the uterus. Cervical cancer develops when cells in the cervix begin to multiply abnormally. There are over 20 serotypes of HPV that affect the genital areas. Harro et al., who investigated the safety and immunogenicity of a virus-like particle papillomavirus vaccine, define seropositive as an ELISA antibody titre greater than or equal to the reactivity of a standard pooled serum [4]. At study start, 6 out of the 72 females were seropositive, and seroconversion was defined as a fourfold or greater rise in titre.

Yet another definition of seroconversion is one that combines the two given above: becoming seropositive if seronegative at enrolment, or a fourfold rise if seropositive at enrolment. For example, in clinical influenza vaccine studies seroconversion is usually defined as: an anti-HA antibody titre <10 at baseline and a post-vaccination titre ≥ 40 *or* a titre >10 at baseline and at least a fourfold increase in titre post-vaccination. (In fact, the reader may note that this is a third alternative definition, because the definition thus not say: a baseline titre <10 (= seronegative) and a post-vaccination titre ≥ 10 (= seropositive), but, a post-vaccination titre ≥ 40 (= seroprotected).)

Whichever of the above definitions is used, just like seroprotection, seroconversion is a binary endpoint, and the *seroconversion rate*—the fraction of study subjects who seroconverted—is a rate (i.e. a proportion, see Notation).

3.5 Analysis of Rates

In this section, the analysis of seroprotection and seroconversion rates is discussed. However, this will be done in the wider context of analysing and comparing rates. The reason for doing so is that in clinical vaccine trials binary endpoints and thus rates are very common: rates of subjects reporting local or systemic reactions, rates of subjects reporting a particular adverse vaccine event, rates of subjects remaining disease-free after vaccination, etc. Thus, the methods discussed in this section are not only applied to seroprotection and seroconversion rates but also to many other kinds of rates.

Two types of tests are discussed, exact and asymptotic. The actual coverage rate of exact confidence intervals is usually not $100(1-\alpha)\%$ (as it should be), but *at least* $100(1-\alpha)\%$. And because exact confidence intervals tend to be too wide, they will include the null value more often than intervals that are not too wide. This implies that if the null hypothesis is true, the probability that it is rejected is not equal to the significance level but below it. This may look like an advantage, but it is not because there is a price to pay. Because a too wide interval may include the null value, there is a negative and thus undesirable impact on the statistical power when the alternative hypothesis is true. And this is why it is said that exact tests are conservative. However, not all exact tests are equally conservative. Barnard's test,[1] for example, is less conservative than Fisher's exact test. Asymptotic tests are not conservative (at least, when no continuity correction is applied), but they require non-rare rates and non-small sample sizes. For this reason, in the analysis of safety data (see Chap. 12), where events are often rare to very rare, many prefer the use of exact tests, despite their conservatism. As an example of such an exact analysis, see Example 12.2.

3.5.1 *Analysis of a Single Rate*

Null hypotheses about the rate π of a particular binary event, for example, becoming seroprotected or having seroconverted, can be statistically tested using a test based on the binomial distribution $B(n,\pi)$, where n is the number of observations. To test the null hypothesis H_0: $\pi \leq \pi_0$ against the one-sided hypothesis H_1: $\pi > \pi_0$ the tail probability $\Pr(\mathbf{X} \geq c|\pi_0)$ is calculated, where $\mathbf{X}$ is a $\text{Bin}(n, \pi_0)$ distributed random variable and c the observed number of events. This probability can be found with the SAS function PROBBNML(p, m, k), which returns the probability that an observation from a BIN(m, p) distribution is less or equal to k.

Example 3.4 Feiring et al. report the results of a study with a meningococcal B vaccine in a group of 374 children, of whom 248 were randomized to the meningococcal B vaccine group and 126 to the placebo group [5]. Antibodies were measured with the serum bactericidal assay (SBA), and seroprotection was defined as a SBA titre $\geq$4. In total, 226 children received all three doses of the meningococcal B vaccine. Six weeks follow-up immunogenicity data were available for 218 children, of whom 132 were seroprotected at follow-up. Assume that the null hypothesis is that $\pi = 0.5$. The observed seroprotection rate is $132/218 = 0.606$, and

$$\begin{aligned}\Pr(\mathbf{X} \geq 132|0.5) &= 1 - \text{PROBBNML}(0.5,\ 218,\ 131)\\ &= 1 - 0.999 = 0.001.\end{aligned}$$

Thus, if the null hypothesis is tested at the one-sided significance level 0.025, it can be rejected.

More common than testing a null hypothesis about a rate π is to calculate a confidence interval for it.

[1] Suissa–Shuster test in the first edition.

Confidence Intervals for a Single Rate

There are numerous methods to calculate a confidence interval for a single rate π. Some methods are exact, others asymptotic. There is no single superior method. Often used criteria for the evaluation of the different methods are the coverage probability and the expected width of the interval. For a comparison of seven standard methods, the reader is referred to the paper by Newcombe [6]. Here, three methods will be discussed: the Clopper–Pearson method, the Wald method and the Wilson method.

The Clopper–Pearson method is an exact method, based on the binomial test. The lower limit of the $100(1-\alpha)\%$ Clopper–Pearson confidence interval is the largest value for π such that

$$\Pr(\mathbf{X} \geq c|\pi) \leq \alpha/2.$$

Conversely, the upper limit of the interval is the smallest value for π such that

$$\Pr(\mathbf{X} \leq c|\pi) \leq \alpha/2.$$

PROC FREQ of SAS returns Clopper–Pearson confidence limits if requested (use the option BIN EXACT with the TABLES statement).

Also, formula for the limits exists. If $F_{k,l;1-\alpha/2}$ is the $100(1-\alpha/2)$th percentile of the F-distribution with k numerator degrees of freedom and l denominator degrees of freedom, then the lower Clopper–Pearson confidence limit is

$$LCL_{CP} = \frac{c}{c + (n - c + 1)f_L},$$

where $f_L = F_{2(n-c+1),2c;1-\alpha/2}$. The upper Clopper–Pearson confidence limit is

$$UCL_{CP} = \frac{(c + 1)f_U}{(n - c) + (c + 1)f_U},$$

where $f_U = F_{2(c+1),2(n-c);1-\alpha/2}$.

Example 3.4 (continued) In the study, 132 out of 218 children were seroprotected. Upper percentiles of F-distributions can be obtained with the SAS function FINV. The upper percentile is

$$\begin{aligned} f_L &= \text{FINV}(0.975, 2 \times (218 - 132 + 1), 2 \times 132) \\ &= \text{FINV}(0.975, 174, 264) \\ &= 1.307. \end{aligned}$$

So that

$$LCL_{CP} = \frac{132}{132 + (218 - 132 + 1)1.307} = 0.537.$$

Similarly, the upper percentile is

$$\begin{aligned} f_U &= \text{FINV}(0.975, 2 \times (132+1), 2 \times (218-132)) \\ &= \text{FINV}(0.975, 266, 172) \\ &= 1.318 \end{aligned}$$

and

$$UCL_{\text{CP}} = \frac{(132+1)1.318}{(218-132)+(132+1)1.318} = 0.671.$$

Thus, the two-sided 95% Clopper–Pearson confidence interval for the probability of being seroprotected is (0.537,0.671).

A drawback of the Clopper–Pearson method is that it conservative in the sense that the coverage probability of the interval is at least the nominal value, $1-\alpha$.

The Wald method and the Wilson method are both asymptotic methods. The Wald method is the simpler of the two:

$$LCL_{\text{Wald}},\ UCL_{\text{Wald}} = R \pm z_{1-\alpha/2} SE(R),$$

where $R = c/n$, the observed event rate and

$$SE(R) = \sqrt{R(1-R)/n}$$

its estimated standard error. The coverage probability of the Wald interval is on average too low and may be very low if π is in the vicinity of zero or one. When the continuity correction is used the coverage probability is improved but for extreme π it may still be far of the nominal value. In vaccine development, this is a serious drawback because seroprotection rates, for example, are often close to one, while adverse vaccine event rates can be close to zero. Wald-type confidence intervals are based on standard errors estimated from the data, see also Sect. 3.5.2.

An asymptotic method with an average coverage probability close to the nominal value is the Wilson method. Wilson-type confidence intervals are based on standard errors which take the null values of the parameter of interest into account, and are not or only partially estimated from the data, see again Sect. 3.5.2. Under the null hypothesis H_0: $\pi = \pi_0$ the statistic

$$Z = \frac{R - \pi_0}{SE_0(R)}$$

is approximately standard normally distributed, with

$$SE_0(R) = \sqrt{\pi_0(1-\pi_0)/n}$$

the standard error of R under the null hypothesis. According to one of the first principles of statistics, the range of all values for π_0 that are not rejected at the significance level α constitute a $100(1-\alpha)\%$ confidence interval for π. To find the limits of this range, the following equation must be solved

$$\frac{(R-\pi_0)^2}{\pi_0(1-\pi_0)/n} = z_{1-\alpha/2}^2.$$

This leads to the following limits:

$$LCL_{\text{Wilson}}, UCL_{\text{Wilson}} = \frac{-B \pm \sqrt{B^2 - 4AC}}{2A},$$

where

$$A = n + z_{1-\alpha/2}^2, \quad B = -(2s + z_{1-\alpha/2}^2), \quad C = s^2/n.$$

Example 3.4 (continued) With $z_{0.975} = 1.96$, it follows that

$$A = (218 + 1.96^2) = 221.84$$
$$B = -(2 \times 132 + 1.96^2) = -267.84$$
$$C = 132^2/218 = 79.93.$$

Thus, asymptotic two-sided 95% confidence limits for π are

$$LCL_{\text{Wilson}} = \frac{267.84 - \sqrt{267.84^2 - 4 \times 221.84 \times 79.93}}{2 \times 221.84} = 0.539$$

and

$$UCL_{\text{Wilson}} = \frac{267.84 + \sqrt{267.84^2 - 4 \times 221.84 \times 79.93}}{2 \times 221.84} = 0.668.$$

For the example, the confidence limits based on the Wilson method are almost identical to the limits based on the Clopper–Pearson method. These Wilson-type confidence limits can be requested in PROC FREQ by using the option BIN (WILSON) with the TABLES statement.

3.5.2 Comparing Two Rates

There are two statistics to compare two rates R_1 and R_0. These statistics are the rate difference

$$RD = R_1 - R_0,$$

and the rate ratio

$$RR = R_1/R_0.$$

The rate difference is an estimator of the risk difference

$$\Delta = \pi_1 - \pi_0$$

and the rate ratio is an estimator of the relative risk

$$\theta = \pi_1/\pi_0.$$

('Risk' is used here for convenience, although the event need not to be negative, like seroprotection, for example.)

To test the null hypothesis that $\Delta = 0$ is equivalent to testing that $\theta = 1$. Both null hypotheses can be tested with Pearson's chi-square test or, in case of small sample sizes, an exact test, either the well-known Fisher's exact test, or, preferably, the less well-known Barnard's test. This latter test is discussed in Sect. 3.5.3.

To calculate confidence intervals for Δ or θ, both asymptotic and exact methods are available. For both parameters, two types of asymptotic intervals that can be found in the statistical and epidemiological literature are the familiar Wald-type intervals and the less familiar Wilson-type intervals. The Wilson-type intervals should be the intervals of choice because their coverage is superior to that of the Wald-type intervals.

Asymptotic confidence limits for a risk difference Δ are often calculated as

$$RD \pm z_{1-\alpha/2} SE(RD), \tag{3.4}$$

where

$$SE(RD) = \sqrt{R_1(1-R_1)/n_1 + R_0(1-R_0)/n_0}, \tag{3.5}$$

and n_1 and n_0 the sizes of the two groups being compared. These Wald-type confidence limits are available in PROC FREQ (use the option RISKDIFF with the TABLES statement). The Wald approach may mean disagreement between Pearson's chi-square statistic to test the null hypothesis H_0: $\Delta = 0$ and the confidence limits. This is because the test statistic is calculated under the null hypothesis, with the standard error of RD also estimated under the null hypothesis:

$$SE_0(RD) = \sqrt{R(1-R)(1/n_1 + 1/n_0)}, \tag{3.6}$$

where

$$R = \frac{c_1 + c_0}{n_1 + n_0},$$

and c_1 and c_0 the observed numbers of events. Then

$$\chi^2_{\text{Pearson}} = \frac{RD^2}{SE_0^2(RD)}.$$

The Pearson statistic and the Wald-type confidence interval are thus based on different estimates of the standard error of the rate difference—$SE_0(RD)$ versus $SE(RD)$—which explains the occasional disagreement between the test statistic and the interval. Another drawback of the Wald approach is that $SE(RD)$ cannot be calculated if either R_1 or R_0 equals zero or one.

An approach that does not suffer from these drawbacks was proposed by Miettinen and Nurminen [7]. In their approach, based on the Wilson method, the limits of the two-sided 100(1−α)% confidence interval for Δ are those values Δ_0 that satisfy the equation

$$\frac{RD - \Delta_0}{SE_{\Delta_0}(RD)} = \pm z_{1-\alpha/2}, \tag{3.7}$$

where

$$SE_{\Delta_0}(RD) = \sqrt{\tilde{R}_1(1-\tilde{R}_1)/n_1 + \tilde{R}_0(1-\tilde{R}_0)/n_0}. \tag{3.8}$$

$\tilde{R}_1$ and $\tilde{R}_0$ are constrained maximum likelihood estimates of π_1 and π_0, with as constraint

$$\tilde{R}_1 - \tilde{R}_0 = \Delta_0.$$

Miettinen and Nurminen give a closed-formed solution for $\tilde{R}_0$, which is reproduced in Appendix B of this book. The confidence limits have to be found iteratively. A simple iterative approach is the following. If 95% confidence limits are required, with a precision of, say, three decimals, then, to find the upper limit of the interval, evaluate the test statistic on the left-hand side of Eq. (3.7) for $\Delta_0 = RD + 0.001$, $\Delta_0 = RD + 0.002$, etc., until the test statistic exceeds 1.96. The upper limit is the largest tested value for Δ for which the test statistic is less than 1.96. To find the lower limit, evaluate the test statistic for $\Delta_0 = RD - 0.001$, $\Delta_0 = RD - 0.002$, etc., until the test statistic falls below −1.96. The lower limit is the smallest value for Δ for which the test statistic is greater than −1.96. (See Appendix I for a SAS code. Also available in PROC FREQ, use the option RISKDIFF (CL=MN) with the TABLES statement.)

Example 3.5 Consider a randomized immunogenicity trial in which both seroprotection rates are equal to 1.0, say, $r_1 = 48/48$ and $r_0 = 52/52$. The Wald approach does not allow calculation of a 95% confidence interval for Δ, but the Wilson approach does: (−0.074, 0.068).

The standard approach to calculate an asymptotic confidence interval for relative risk θ is to calculate Wald-type confidence limits based on the log-transformed rate ratio RR, which are then back-transformed, the so-called logit limits. In this approach, the standard error of $\log RR$ is estimated as

$$SE(\log RR) = \sqrt{1/c_1 - 1/n_1 + 1/c_0 - 1/n_0}. \tag{3.9}$$

The logit limits of the two-sided 100(1−α)% Wald-type confidence interval for θ are

$$\exp[\log RR \pm z_{1-\alpha/2} SE(\log RR)]. \tag{3.10}$$

These Wald-type confidence limits for θ are also available in PROC FREQ (use the option RELRISK with the TABLES statement).

For the relative risk too, Miettinen and Nurminen derive a Wilson-type confidence interval. The approach is the same as for the rate difference. The limits of the two-sided $100(1-\alpha)\%$ Wilson-type confidence interval for θ are the values θ_0 that satisfy the equation

$$\frac{R_1 - \theta_0 R_0}{SE_{\theta_0}(R_1 - \theta_0 R_0)} = \pm z_{1-\alpha/2}, \tag{3.11}$$

where

$$SE_{\theta_0}(R_1 - \theta_0 R_0) = \sqrt{\tilde{R}_1(1-\tilde{R}_1)/n_1 + \theta_0^2 \tilde{R}_0(1-\tilde{R}_0)/n_0}. \tag{3.12}$$

Again, $\tilde{R}_1$ and $\tilde{R}_0$ are constrained MLEs of π_1 and π_0, with as constraint

$$\tilde{R}_1 = \theta_0 \tilde{R}_0.$$

For the closed-form solution for $\tilde{R}_0$, see Appendix B.

Example 3.5 (continued) Because $SE(\log RR) = 0.0$, the Wald approach does not allow calculation of a confidence interval for θ. The two-sided 95% Wilson-type confidence interval for θ is $(0.926, 1.073)$.

(See Appendix I for a SAS code, not available in PROC FREQ.) The Wilson-type confidence interval for θ cannot be evaluated when either c_1 or c_0 equals zero.

The estimator RR is biased, because it overestimates θ. The explanation is that it is non-linear in the maximum likelihood estimator R_0. The bias will be non-negligible when the control rate θ_0 approaches zero and n_0 is small to intermediate [8]. This may be the case in vaccine field studies (see Chaps. 8 and 9), with often very low attack rates, and in safety analyses (see Chap. 12), when comparing adverse events rates between vaccine groups. At least two bias corrections exist [8]. The simplest but very effective bias correction for the rate ratio is to add one imaginary event to the control group, i.e., to set c_0 to (c_0+1) and n_0 to (n_0+1). This correction is known as Jewell's correction. Somewhat surprisingly perhaps, Jewell's correction does not improve the performance of the Wilson-type confidence interval for the relative risk, in the sense that it would give better coverage. In fact, the uncorrected confidence procedure provides near nominal coverage while the corrected procedure can give subnominal coverage.

3.5.3 Barnard's Exact Test for Comparing Two Rates

Fisher's test is an exact test for comparing two rates, based on conditioning on both the margins of the 2 × 2 table. An alternative to this test is a test based on the maximization method, Barnard's test, which is also an exact test [9]. This test is conditional on the sample sizes being fixed, but it does not condition on the number of observed cases. The test has an attractive property, which Fisher's test lacks, namely, that it can be used to test hypotheses of the form H_0: $\Delta = \Delta_0$, with $\Delta_0 \neq 0$, or H_0: $\theta = \theta_0$, with $\theta_0 \neq 1$. This means that the test can be used to derive exact confidence intervals for risk differences and relative risks.

To calculate the P-value for Barnard's exact test, the procedure proposed by Chan is given [10]. The procedure consists of three steps.

Step 1 is that the appropriate Z statistic to compare two numbers of events is identified. To test the null hypothesis H_0: $\Delta = \Delta_0$ the Z statistic is

$$Z = \frac{RD - \Delta_0}{SE_{\Delta_0}(RD)}, \tag{3.13}$$

where $SE_{\Delta_0}(RD)$ is the standard error (3.8). When the null hypothesis is H_0: $\theta = \theta_0$, the Z statistic is

$$Z = \frac{r_1 - \theta_0 r_0}{SE_{\theta_0}(r_1 - \theta_0 r_0)}, \tag{3.14}$$

with $SE_{\theta_0}(r_1 - \theta_0 r0)$ the standard error (3.12). For $\Delta_0 = 0$ and $\theta_0 = 1$, both statistics reduce to the standard Z statistic to compare two rates:

$$Z = \frac{RD}{SE_0(RD)}, \tag{3.15}$$

with $SE_0(RD)$ the standard error (3.6). Let Z_{obs} be the value for the appropriate Z statistic for the observed numbers of events (c_1, c_0).

Step 2 is that for every possible combination (i, j) of numbers of events the value Z_{ij} for the Z statistic is calculated.

Step 3 is to find a P-value for testing the null hypothesis. When $Z_{\text{obs}} \geq 0$, the one-sided P-value for Barnard's test is defined as

$$P_{\max} = \max_{\{\pi_0 \in D\}} \Pr(Z \geq Z_{\text{obs}} | \pi_0).$$

For a given value for π_0, $\Pr(Z \geq Z_{\text{obs}} | \pi_0)$ is the sum of the probabilities of those 2 × 2 tables with $Z_{ij} \geq Z_{\text{obs}}$. Under the null hypothesis, these probabilities are the products of two binomial probabilities:

$$\binom{n_1}{i} \pi_1^i (1 - \pi_1)^{n_1 - i} \times \binom{n_0}{j} \pi_0^j (1 - \pi_0)^{n_0 - j}$$

with $\pi_1 = (\pi_0 + \Delta_0)$ for the Z statistic (3.13) and $\pi_1 = \theta_0\pi_0$ for the Z statistic (3.14). When *RD* or *RR* is not consistent with the one-sided alternative hypothesis, e.g. when the alternative hypothesis is $\Delta < 0$ but $RD > 0$, the one-sided P-value is $1 - P_{max}$.

The parameter π_0 is a nuisance parameter. The domain for π_0 is the continuous interval

$$D = \begin{cases} (0, 1 - \Delta_0), & \Delta_0 > 0 \\ (-\Delta_0, 1), & \text{if} \quad \Delta_0 < 0 \\ (0, 1/\theta_0), & \theta_0 > 1 \\ (0, 1), & \Delta_0 = 0 \text{ or } \theta_0 \leq 1. \end{cases}$$

Chan proposes to divide D in a large number of equally spaced intervals and calculate the probability at every increment, for example (0.001, 0.002, ..., 0.999) if D is (0, 1), an approach which would provide sufficient accuracy for most practical uses. The computer must be instructed to set the Z statistic (3.14) to zero for the 2×2 table with numbers of events (n_1, n_0).

When $Z_{obs} \leq 0$, the one-sided P-value for Barnard's test is defined as

$$P_{max} = \max_{\{\pi_0 \in D\}} \Pr(Z \leq Z_{obs} | \pi_0),$$

or $1 - P_{max}$ in case of an inconsistency between *RD* or *RR* and the alternative hypothesis.

The two-sided P-value for Barnard's test is defined as

$$P_{max} = \max_{\{\pi_0 \in D\}} \Pr(|Z| \geq |Z_{obs}| | \pi_0).$$

Barnard's test for risk difference analysis is available in PROC FREQ.

SAS Code 3.1 Barnard's test for exact risk difference analysis

```
data;
    input Vaccine Illness Count @@;
datalines;
1 1 7 1 0 8 0 1 12 0 0 3
;

proc freq;
    tables Vaccine*Illness;
    exact barnard riskdiff (method=score);
    weight Count;
run;
```

Example 3.6 Chan cites a challenge study on the protective efficacy of a recombinant protein influenza vaccine. In the study 15 vaccinated and 15 placebo subjects were challenged with a weakened A-H1N1 influenza virus strain. After 9 days the observed rates of any clinical illness were 7/15 in the vaccine group and 12/15 in the control group, the placebo group. Here, the null hypothesis $H_0 : \Delta = 0$ is to be tested against the one-sided alternative $H_0 : \Delta < 0$. The observed value for the Z statistic is $Z_{obs} = -1.894$. Barnard's test yields a one-sided P-value of 0.0341 and a two-sided P-value of 0.0682. As exact two-sided 95% confidence interval for Δ (–0.637, 0.239) is found.

In Appendix I, a SAS code for Barnard's test for relative risk analysis is available (SAS Code I.2). An exact confidence interval for a relative risk can be obtained by testing one-sided hypotheses for several different values for θ.

Example 3.6 (continued) For the challenge data, the exact two-sided 95% confidence interval for the relative risk θ is found by trial and error. The one-sided P-value for the null hypothesis H_0: $\theta = 0.260$ is 0.0218, and that for H_0: $\theta = 0.261$ is 0.0251. Thus, the lower limit of the 0.261. The one-sided P-values for the null hypotheses H_0: $\theta = 1.037$ and $H_0 : \theta = 1.038$ are 0.0250 and 0.0248, respectively. Thus, the upper limit of the exact two-sided 95% confidence interval is 1.037.

Lydersen et al. compare the performance of Fisher's exact test and Barnard's and other similar tests [11]. The performance of Barnard's exact test is superior to that of Fisher's exact test, which is conservative. They advise that Fisher's exact test should no longer be used.

3.6 Multiplicity in Immunogenicity Trials and the Intersection-Union Test

When there are multiple serotypes for a given pathogen and a vaccine trial is done in support of marketing approval of the vaccine, the multiplicity issue arises. This multiplicity is of a special nature and therefore deserves emphasis. Usually what is to be demonstrated is an improvement in immunogenicity for *all* serotypes included in the vaccine. Prevnar 13 is a vaccine to prevent infection caused by pneumococcal bacteria. As indicated by the name it is a 13-valent vaccine, and for its approval statistical significance had to be achieved for *all* 13 included *Streptococcus pneumoniae* strains. For this scenario, when significance must be achieved for all serotypes, a much applied approach is the one based on the intersection-union (IU) test [12]. In this approach, for each of the k co-primary endpoints, the component null hypothesis is tested at the significance level α, and superiority of the investigational vaccine to the control vaccine is claimed only if all k component null hypotheses are rejected. There is no alpha penalty. On the plus side of the IU test is its simplicity, which makes the test easy to explain to non-statisticians. On the negative side is that the test is known to be conservative. Only when the k endpoints are perfectly correlated the significance level of the test will be exactly α, in all other cases the level will be less than α. When the k endpoints are independent, the level of the test will be as low as α^k.

A similar scenario occurs when significance must be achieved for more than one immunogenicity endpoint, for example, for both the seroprotection and the seroconversion rate, in which case there are two co-primary endpoints (for monovalent vaccines), or four co-primary endpoints (for bivalent vaccines), etc.

Several alternatives to the IU test have been proposed. For the case that the data follow a multivariate normal distribution, for example, Laska et al. show that under mild conditions a test known as the min test is uniformly the most powerful test [13]. The min test statistic Z_{min} is defined as

$$Z_{\text{min}} = \min\{Z_1, \ldots, Z_k\},$$

where Z_i is a test statistic to test the ith component null hypothesis. The sampling distribution of the test statistic is, however, complicated and depends amongst others on the covariance structure of the endpoints, which limits the applicability of the test.

3.7 The Reverse Cumulative Distribution Plot

The *reverse cumulative distribution plot* is a graphic tool to display the distribution of immunogenicity values. It is particularly useful for visual comparisons of distributions between vaccine groups. The plot became quickly popular after a lucid presentation of its properties by Reed et al. [14].

In Fig. 3.1, four examples of a reverse cumulative distribution (RCD) curve are shown. The x-axis represents the immunogenicity values, and the scale of the axis is usually logarithmic. The y-axis represents the percentage of subjects having at least that immunogenicity value. Thus, to a value x on the x-axis corresponds the percentage of subjects having a log-transformed immunogenicity value greater or equal to x. By definition, the curve begins at 100%, and then descends down, from left to right. The lowest point on the curve is the percentage of subjects having a log-transformed immunogenicity value equal to the largest observed value. The plot is called the reverse cumulative distribution plot because it reverses the cumulative distribution. The median log-transformed immunogenicity level is the value on the x-axis below the y-axis value of 50%.

If the distribution of the log-transformed immunogenicity values is symmetric with little variability around the mean, then the middle section of the RCD curve will be steep (curve A). If, on the other hand, the variability is large then the middle section of the curve will be less steep (curve B). In the extreme, if the distribution of the log-transformed immunogenicity values is more or less uniform, then the curve will be approximately a downward-sloping straight line. If the distribution of the log-transformed immunogenicity values is skewed to the right, with a large fraction

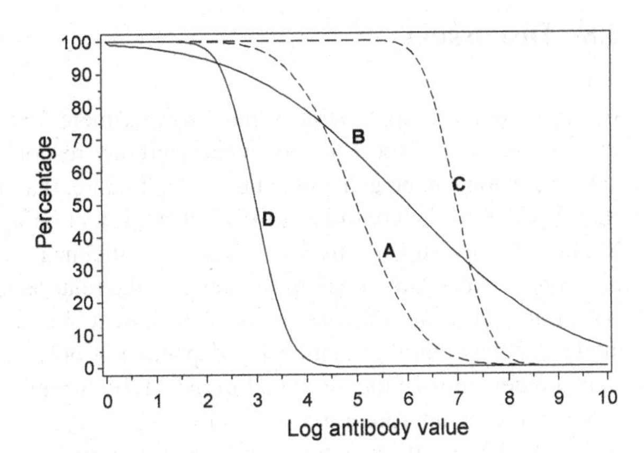

Fig. 3.1 Four examples of a RCD curve

of the subjects having a value near the high end, then the curve will be rectangular, i.e. remaining high and flat with a rapid descend near the end (curve C). If the distribution of the log-transformed immunogenicity values is skewed to the left, thus if there are many subjects with a low value, then the curve will be similar to curve D.

If the RCD curve for one vaccine is above the curve for another vaccine, and the two curves do not intersect (e.g. A and D in Fig. 3.1), then every percentile of distribution of the immunogenicity values of vaccine A is higher than the corresponding percentile for vaccine D, meaning that vaccine A induced the higher immune responses. If the two curves intersect, then one vaccine induced both more lower and higher immune responses. If (the y-value of) the point of intersection is above 50% (as for B and A in Fig. 3.1), then this is in favour of vaccine B, because compared to group A, in group B there would be a larger fraction of subjects with a high immune response, while a point of intersection below 50% would be in favour of vaccine A.

There is a similarity between RCD curves and survival curves. Both curves display percentages of subjects having a value (either an immunogenicity value or a survival time) greater or equal the values on the x-axis. From this similarity, Small et al. conclude that RCD curves can be compared between vaccine groups using rank-based tests for survival analysis, such as the log-rank test or the generalized Wilcoxon test [15]. Because these tests are non-parametric they do not require normally distributed data, they are an interesting alternative to the geometric mean titres analysis discussed in Sect. 3.2.2. It is well-known that when the distribution of the data is known, parametric tests and rank-based tests do not differ strongly in power. When, as often is the case, two RCD curves are separated in the middle but not in the tails (e.g. curves A and C in Fig. 3.1), then the generalized Wilcoxon test is more powerful than the log-rank test. On the other hand, when two curves are separated both in the middle and the tails (without intersecting), then the log-rank test is the more powerful of the two.

3.8 Discussion

When it has been demonstrated that an experimental vaccine is superior to a control vaccine with respect to, say, mean antibody response, it is often assumed that the conclusion can be generalized to protection, i.e. that the experimental vaccine is also superior to the control with respect to protection against disease or infection. Nauta et al. investigate this for the case of influenza vaccines [16]. With the help of a simple statistical model, they show that the relationship between antibody level and protection from influenza is more complicated than perhaps envisioned. Their model predicts that the relationship depends not only on the mean but also on the standard deviation of the log-transformed antibody values. It is this dependency that complicates the interpretation. They observe, for example, that if the mean antibody level of both the experimental and the control group are high, a positive difference in mean level implies a positive difference in the fraction protected subjects, unless the difference in mean level is small to moderate in combination with a large, positive difference in standard deviation. Interpretation of differences in fractions seroprotected subjects is even more challenging. Their model predicts that differences in the fraction seroprotected subjects cannot be interpreted without taking into account the mean antibody levels and standard deviations, shedding a new and unexpected light on the usefulness of the concept of seroprotection.

It is not unreasonable to assume that these observations also hold for many other vaccines. The implication would be that the standard methods discussed in this chapter are in need of improvement in the sense that they compare means rather than (simultaneously) means and standard deviations of distributions. An early attempt in this direction is made by Lachenbruch et al., who outline a statistical method based on measuring the similarity between the scales and shapes of antibody level distributions [17]. Computationally, their method is complex. For example, critical values for the test statistic have to be simulated.

3.9 Sample Size Estimation

3.9.1 Comparing Two Geometric Mean

There are a number of formulae for sample size estimation for parallel group trials with normal data. The simplest but least accurate formula, which can be found in any basic text on statistics, is the well-known formula based on a normal approximation. The most accurate one, involving no approximations, is based on a non-central t-distribution, see formula (3.8) in the book by Julious [18]. Sample sizes based on this formula can be obtained with PROC POWER. Two parameters have to be specified, the difference Δ of the log-transformed geometric means and the within-group standard deviation σ of the log-transformed immunogenicity values. The value for σ may be taken from previous trials. Alternatively, the following formula can be used to obtain a, conservative, value for σ:

$$\sigma \approx \frac{\log(\text{largest value}) - \log(\text{smallest value})}{4}.$$

This formula is based on the fact that in case of normal data approximately 95% of the observations will fall in the range $\mu - 2\sigma$ to $\mu + 2\sigma$. To obtain a less conservative value, in the formula above the multiplier 4 should be substituted with 5 or 6.

Example 3.7 Consider a clinical trial in which two influenza vaccines are to be compared, a licensed one and a new, investigational vaccine, with the antibody response as measured by the HI test as primary endpoint. Assume that the investigator believes that the new vaccine is considerably more immunogenic than the licensed one, and that the geometric mean ratio will be ≥ 2.0. In influenza vaccine trials HI titres $>5{,}120$ are rare, and the lowest possible value is usually 5, i.e. half of the reciprocal of the starting dilution, 10. Thus

$$\Delta = \log 2.0 = 0.69,$$

and a conservative value for σ is

$$\sigma = \frac{\log 5{,}120 - \log 5}{4} = 1.73.$$

Below the SAS code to estimate the number of subjects per vaccine group for a statistical power of 0.9 is given. The required sample size is found to be 134 subjects per group, 268 subjects in total.

SAS Code 3.2 Sample size calculation for comparing two geometric means

```
proc power;
   twosamplemeans test=diff
   meandiff=0.69 stddev=1.73
   power=0.9 npergroup=.;
run;
```

For the reverse approach, finding the statistical power for a given sample size, in the SAS code POWER should be set to missing (.) and NPERGROUP to the proposed number of subjects per group.

3.9.2 Comparing Two Rates

For the estimation of the sample size required to compare two rates also, numerous formulae exist. All these formulae are asymptotic, and a detailed discussion of them can be found in chapter 4 of the book by Fleiss, Levin and Paik [19]. The formulae they advise, (4.14) for equal sample sizes and (4.19) for unequal sample sizes, have been included in the PROC POWER. For a discussion on exact power calculations for 2×2 tables, the reader is referred to the paper by Hirji et al. [20].

Example 3.7 (continued) Assume that the investigator wants to compare seroprotection rates rather than geometric mean titres, and that he expects an increase in the seroprotection rate from 0.85 to 0.95. Below the SAS code to calculate the number of subjects per vaccine group for a statistical power of 0.9 is given. The required sample size is found to be 188 subjects per group, see SAS Code 3.3.

SAS Code 3.3 Sample size calculation for comparing two rates

```
proc power;
   twosamplefreq test=pchi
   groupproportions= (0.85,0.95)
   power=0.9 npergroup=.;
run;
```

3.9.3 *Sample Size Estimation for Trials with Multiple Co-primary Endpoints*

Estimation of the power of a trial with multiple co-primary endpoints is a non-trivial problem. The key of the problem is that the power is critically dependent on the correlation between the endpoints. Different assumptions about the correlation can lead to substantially different sample size estimates. The power will be the highest when the endpoints are strongly correlated but will be lowest when the endpoints are minimally correlated. In practice, the correlations between the endpoints will often be unknown. If the objective of the trial is to demonstrate statistical significance for all k co-primary endpoints (see Sect. 3.6), an often applied approach is to obtain a lower bound for the global power P using the following inequality:

$$P \geq \prod_{i=1}^{k} P_i, \tag{3.16}$$

with P_i the power of the trial for the ith endpoint. This inequality requires the assumption that all endpoints are non-negatively correlated.

An inequality that does not require any assumptions about the correlations is

$$P \geq \sum_{i=1}^{k} P_i - (k-1). \quad (3.17)$$

(See Appendix C for a proof of this inequality.) When k is large, both inequalities require that the P_i's must be close to 1.0 to be secured of a lower bound exceeding 0.8.

Example 3.8 Consider a comparative trial with two co-primary endpoints, the geometric mean concentration and the seroprotection rate. Assume that a sample size of 2×150 subjects is being considered, and that with the help of the SAS codes in the previous sections the power for the first co-primary endpoint has been estimated as 0.93 and for the second co-primary endpoint as 0.90. Under the (reasonable) assumption that the two endpoints are non-negatively correlated, a lower bound for the global statistical power is

$$P \geq 0.93 \times 0.90 = 0.837.$$

If inequality (3.17) is used instead, the lower bound for the global power is

$$0.93 + 0.90 - 1 = 0.830.$$

An alternative method to estimate the global power is Monte Carlo simulation. As a simple example, consider an open study with the aim to demonstrate the immunogenicity of a new formulation of a particular vaccine, by showing that both the seroprotection and the seroconversion rate exceed predefined bounds. A large number of studies—minimally 5,000—is simulated. Per simulated study, a random draw with replacement of size n from the database is made, and the result of the study is considered significant if the observed seroprotection and seroconversion rate exceed the predefined bounds. An estimate of the global power of the design for sample size n is the fraction of simulated studies with a significant result.

Chapter 4
Antibody Titres and Two Types of Bias

Abstract In this chapter, two types of possible bias for antibody titres are explored. The first type of bias is due to how antibody titres are defined. An alternative definition is presented, the mid-value definition. With this definition, the bias is properly corrected. The second type of bias occurs when antibody titres above (or below) a certain level are not determined. If this bias is ignored, the geometric mean titre will be underestimated. It is demonstrated how the method of maximum likelihood estimation for censored observation can be applied to eliminate this bias.

4.1 Standard Antibody Titres Versus Mid-Value Titres

Statisticians have pointed out that when antibody titres are determined using the standard definition—the reciprocal of the highest dilution at which the assay read-out did occur (see Sect. 2.1.2)—, the true titre value is underestimated [21]. It is easy to see why. By definition, the true antibody titre τ lies between the standard titre t_s and the reciprocal of the next dilution, r_n:

$$t_s \leq \tau < r_n.$$

Thus, for most serum samples the standard antibody titre will be lower than the true titre. This means that if standard antibody titres are used, the geometric mean titre will underestimate the geometric mean of the distribution underlying the antibody values.

There are antibody assays that try to correct for this bias. An example is the interpolated serum bactericidal assay (SBA) to demonstrate humoral immune responses induced by meningococcal vaccines. Meningococcal disease is caused by the bacterium *Neisseria meningitidis*, also known as *meningococcus* bacteria. Attack rates of the disease are highest among infants aged younger than two years and adolescents between 11 and 19 years of age. The disease can cause substantial mortality. Five serogroups, A, B, C, Y and W135, are responsible for virtually all cases of the disease. The standard SBA titre is defined as the reciprocal of the highest dilution of serum immediately preceding the 50% survival/kill value for colony-forming units

J. Nauta, *Statistics in Clinical and Observational Vaccine Studies*,
Springer Series in Pharmaceutical Statistics,
https://doi.org/10.1007/978-3-030-37693-2_4

(50% cut-off). The interpolated SBA titre is calculated using a formula that determines the percentage kill in dilutions on either side of the 50% cut-off. The titre is the reciprocal of the dilution of serum at the point where the antibody curve intersects the 50% cut-off line.

Another approach to correct for the bias is changing the definition of the antibody titre to

$$\text{antibody titre} = \sqrt{t_s \times r_n},$$

the geometric mean of the standard antibody titre and the reciprocal of the next dilution. For titres based on serial twofold dilutions r_n equals $2 \times t_s$, in which case the definition of the antibody titre becomes

$$\text{antibody titre} = \sqrt{2} \times t_s.$$

This definition is called the *mid-value definition* of antibody titres [22]. Mid-value antibody titres are higher than standard titres. For example, if the predefined dilutions are 1:4, 1:8, 1:16, etc., and the standard antibody titre for a serum sample is 64, then the mid-value titre for the sample is the geometric mean of 64 and 128

$$\sqrt{64 \times 128} = 90.5 = \sqrt{2} \times 64.$$

On a logarithmic scale, the mid-value antibody titre t_{mv} is the mid-point between t_s and r_n:

$$\log t_{mv} = (\log t_s + \log r_n)/2.$$

Hence the name mid-value antibody titre.

In almost all practical situations the mid-value definition reduces the bias of standard antibody titres, meaning that on average the mid-value titre is closer to the true titre than the standard titre, i.e.

$$|t_{mv} - \tau| \leq |t_s - \tau|.$$

A sufficient set of conditions for the mid-value definition to reduce the bias is:

1. The log-transformed antibody titres are normally distributed.
2. The dilutions are predefined.
3. The distance between two consecutive log-transformed dilution factors is small compared to the range of observed log-transformed titre values.

For post-vaccination titres, this is usually the case. (For pre-vaccination titres, often condition 1 or 3 is not met.)

In case of serial twofold dilutions, when the antibody titres are determined using the standard definition and the primary outcome measure is the geometric mean titre, there is no need to calculate the individual mid-value titres. Let GMT_s be the geometric mean of a single group of standard antibody titres. Then the geometric mean of the corresponding mid-value titres is

$$GMT_{mv} = \sqrt{2} \times GMT_s.$$

A similar expression holds for the confidence limits for GMT_{mv}.

For the geometric mean fold increase, which can be expressed as the ratio of a post- and a pre-vaccination geometric mean titre (see Sect. 3.3.1), the bias correction is also not needed, because

$$gMFI_{mv} = \frac{GMT_{mv}\ \text{post}}{GMT_{mv}\ \text{pre}} = \frac{\sqrt{2}GMT_s\ \text{post}}{\sqrt{2}GMT_s\ \text{pre}}$$
$$= \frac{GMT_s\ \text{post}}{GMT_s\ \text{pre}} = gMFI_s.$$

When the summary statistic of interest is the geometric mean fold increase, both definition, the standard and the mid-value one, will produce the same result. The same holds true for the geometric mean ratio, the ratio of two independent geometric mean titres:

$$GMR_{mv} = \frac{GMT_{mv\ 1}}{GMT_{mv\ 0}} = \frac{\sqrt{2}GMT_{s\ 1}}{\sqrt{2}GMT_{s\ 0}}$$
$$= \frac{GMT_{s\ 1}}{GMT_{s\ 0}} = GMR_s.$$

Both definitions produce the same result.

4.2 Censored Antibody Titres and Maximum Likelihood Estimation

If the number of dilutions in an antibody assay is limited, then it may happen that at the highest tested dilution the assay read-out did not occur. In that case, often the titre is set to the reciprocal of the highest dilution. The result of this practice is bias. The geometric mean titre will be underestimated, and the assumption of normality for the distribution of the log-transformed titres may not hold.

Example 4.1 In Fig. 4.1, the histogram of the frequency distribution of log-transformed post-vaccination measles HI antibody titres of a hypothetical group of 300 children is displayed. The starting dilution was 1:4, and as log transformation the standard transformation ($\log_2$(titre/2)) was used. The arithmetic mean of the log-transformed titres is 5.80, with estimated standard deviation 2.49. This arithmetic mean corresponds to a geometric mean titre of

$$2 \times 2^{5.80} = 111.4.$$

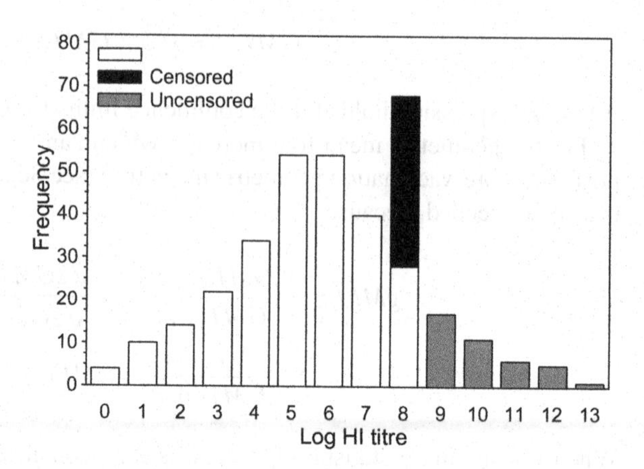

Fig. 4.1 Uncensored and censored frequency distribution of log-transformed measles HI antibody titres

The median and the maximum of the titres are 128 and 16,384, respectively. Next, assume that the highest tested dilution would have been 1:512 rather than a much higher dilution, and that titres above 512 would have been set to 512. The result of this censoring is also shown in Fig. 4.1. The bars above 9, 10, 11, 12 and 13 are added to the bar above 8. Due to the censoring, the frequency distribution is no longer symmetrical and thus no longer normally shaped. The arithmetic mean of the censored log-transformed titres is 5.53 ($GMT = 92.4$), with estimated standard deviation 2.05. Both estimates are smaller than the estimates based on the uncensored data.

In the following sections, it is explained how this bias due to censoring can be eliminated [23].

4.2.1 *ML Estimation for Censored Normal Data*

A censored observation is an observation for which a lower or an upper limit is known but not the exact value. An example of a censored observation is a value below the detection limit of a laboratory test. The upper limit of the test result is known, the detection limit, but not the test result itself. If for an observation only an upper limit is known, the observation is called left-censored. Observation for which only a lower limit is known is called right-censored. Censored observations are often assigned the value of the limit. If this is not taken into account in the statistical analysis, estimates will be biased. A powerful statistical method to eliminate this bias is maximum likelihood (ML) estimation for censored data. As an introduction to ML estimation for censored antibody titres, in this section ML estimation for censored normal data is discussed.

Let $x_1, \ldots, x_n$ be a group of non-censored $N(\mu, \sigma^2)$ distributed observations. The log-likelihood function is

$$LL(\mu, \sigma) = \sum_{i=1}^{n} \log f(x_i; \mu, \sigma),$$

where $f(x; \mu, \sigma)$ is the normal density function. The ML estimates of μ and σ are those values that maximize the log-likelihood function. For normal data, the ML estimates are the arithmetic mean and the estimated standard deviation (but with the $(n - 1)$ in the denominator replaced by n).

Next, assume that r of the observations are right-censored. Let $x_1, \ldots, x_{n-r}$ be the non-censored observations and $x_{n-r+1}, \ldots, x_n$ the right-censored observations. Then the log-likelihood function becomes

$$LL(\mu, \sigma) = \sum_{i=1}^{n-r} \log f(x_i; \mu, \sigma) + \sum_{i=n-r+1}^{n} \log[1 - F(x_i; \mu, \sigma)],$$

where $F(x; \mu, \sigma)$ is the normal distribution function. Thus, for a censored observation x, the density for x is replaced with the probability of an observation beyond x. Again, the ML estimates of μ and σ are those values that maximize the log-likelihood function, and they are unbiased estimates of these parameters.

Finally, assume that there are l left-censored observations: $x_1, \ldots, x_l$. The log-likelihood function for normal data with both left- and right-censored observations is

$$LL(\mu, \sigma) = \sum_{i=1}^{l} \log F(x_i; \mu, \sigma) + \sum_{i=l+1}^{n-r} \log f(x_i; \mu, \sigma) + \sum_{i=n-r+1}^{n} \log[1 - F(x_i; \mu, \sigma)].$$

Maximum likelihood estimation for censored normal data can be intuitively understood as follows. For a series of values for μ and σ a normal curve is fitted to the histogram of the frequency distribution of the observation, and it is checked how well the curve fits to the data. This includes a comparison of the areas under the left and the right tail of the fitted curve with the areas of the histogram bars below or above the censored values. The censored tails are reconstructed, and correct estimates of the mean and standard deviation of the distribution are obtained. The ML estimates of μ and σ are those values that give the best fit, and they are found by iteration.

Maximum likelihood estimation for censored observations was introduced for the analysis of survival data, where it is used to handle censored survival times [24].

4.2.2 ML Estimation for Censored Antibody Titres

To obtain ML estimates for censored, log-transformed antibody titres, SAS procedure PROC LIFEREG can be used. This is a procedure to fit probability distributions to data sets with censored observations. A wide variety of distributions can be fitted, including the normal distribution.

Before the log-likelihood function for censored observations can be applied to log-transformed antibody titres, a modification is needed. In the previous section, it was assumed that the data were censored normal observations. Log-transformed antibody titres, however, are not continuous observations; they are so-called interval-censored observations. An interval-censored observation is an observation for which the lower and the upper value is known but not the exact value. If this is not taken into account, i.e. if the values are treated as if continuous, the ML estimates PROC LIFEREG returns will be invalid.

In Sect. 4.1, it was explained that the true titre value τ lies between the standard titre t_s and the reciprocal r_n of the next dilution:

$$t_s \leq \tau < r_n.$$

Thus, let t_i be an interval-censored standard titre, with r_i the reciprocal of the next dilution. The second term of the log-likelihood function, the term for the non-censored observations becomes

$$\sum_{i=l+1}^{n-r} \log\left[F(\log r_i;\, \mu, \sigma) - F(\log t_i;\, \mu, \sigma)\right].$$

The first term of the log-likelihood function, the term for the left-censored observations, becomes

$$\sum_{i=1}^{l} \log F(\log r_L;\, \mu, \sigma),$$

where r_L is the reciprocal of the starting dilation. The third term of the log-likelihood function, the term for the right-censored observations, becomes

$$\sum_{i=n-r+1}^{n} \log\left[1 - F(\log r_H;\, \mu, \sigma)\right],$$

where r_H is the reciprocal of the highest dilution tested.

In PROC LIFEREG this modification can be handled by the MODEL statement with the LOWER and UPPER syntax; `Lower` and `Upper` are two variables containing the lower and the upper ranges for the observations. For an interval-censored standard titre

$$\texttt{Lower} = \log t_i$$

and

$$\texttt{Upper} = \log r_i$$

For left-censored standard titres `Lower` has to be set to missing (interpreted by PROC LIFEREG as minus infinity) and `Upper` to $\log r_L$; for right-censored standard titres `Lower` $= \log r_H$ and `Upper` has to be set to missing (interpreted as plus infinity). By definition, all observations below the detection limit are left-censored.

Example 4.2 (*continued*) The starting dilution was 1:4, and thus antibody titres less or equal to 4 are to be considered as left-censored. Below a SAS code to fit a normal distribution to the censored log-transformed antibody titres in Fig. 4.1 is given.

Two parameters are estimated, an intercept, which is the ML estimate of μ, and a scale parameter, which is the ML estimate of σ. For both parameters, two-sided 95% confidence limits are given. Note that the μ estimated is the mean of the distribution underlying the log-transformed mid-value titres (Sect. 4.1), and not the mean of the distribution underlying the log-transformed standard titres. This is due to the values assigned to the SAS variables `Lower` and `Upper`, which are consistent with the mid-value definition. To estimate the μ consistent with the definition of standard titres, in the above SAS code `Lower` should be set to `Logtitre−0.5` and `Upper` to `Logtitre+ 0.5`. Then SAS Output 4.1B is obtained.

SAS Code 4.1 Fitting a normal distribution to the censored antibody titres of Fig. 4.1

```
data;
   input Titre Count @@;
   Logtitre=log(Titre/2)/log(2);
   if (Titre > 4) then Lower=Logtitre;
   if (Titre < 512) then Upper=Logtitre+1;
datalines;
4 14 8 14 16 22 32 34 64 54 128 54 256 40 512 68
;

proc lifereg;
  model (Lower,Upper)= / d=normal;
  weight Count;
run;
```

SAS Output 4.1A

```
                              Standard 95% Confidence Chi-
Parameter DF Estimate  Error    Limits          Square Pr > ChiSq

Intercept 1  6.2270    0.1443   5.9443 6.5098   1863.11     <.0001
Scale     1  2.3918    0.1242   2.1604 2.6480
```

The ML estimate of μ is now consistent with the standard titre definition. (This value could have of course also been obtained by subtracting 0.5 from the ML estimates in SAS Output 4.1A: $6.227 - 0.5 = 5.727$.) Note that the correction does not have an effect on the ML estimate of σ.

SAS Output 4.1B

```
                          Standard  95% Confidence   Chi-
Parameter DF Estimate     Error     Limits           Square Pr > ChiSq

Intercept 1  5.7270       0.1443  5.4443  6.0098  1575.93   <.0001
Scale     1  2.3918       0.1242  2.1604  2.6480
```

The ML estimates 5.73 and 2.39 are in good agreement with the estimates based on the uncensored data, 5.80 and 2.49. This demonstrates the powerful tool ML estimation for censored observations is.

Above as log transformation, the standard transformation $\log t = \log_2[t/(D/2)]$ was used, with D the starting dilution factor. A general SAS code to fit a normal distribution to the censored $\log_e$ transformed serial twofold antibody titres is presented below.

SAS Code 4.2 Fitting a normal distribution to censored serial twofold antibody titres

```
data;
   input Titre;
   Midvalue=1;         /* 1 for mid-value definition, 0 otherwise */
   Rsd=4;              /* reciprocal starting dilution            */
   Rhd=512;            /* reciprocal highest dilution             */

   if (Titre > Rsd) then Lower=(log(Titre) +
                      (1-Midvalue)*log(Titre/2))/(2-Midvalue);
  if (Titre < Rhd) then Upper=(log(2*Titre) +
                      (1-Midvalue)*log(Titre))/(2-Midvalue);
datalines;
 .
 ;

proc lifereg;
  model (Lower,Upper)= / d=normal;
run;
```

The approach discussed above can be readily extended to the case of two vaccine groups. Let `Group` be the SAS variable for the groups, taking the value 1 for the experimental group and 0 for the control group. Then the MODEL statement in SAS Code 4.1 must be changed to

```
proc lifereg;
   model (Lower,Upper)=Group / d=normal;
run;
```

Chapter 5
Adjusting for Imbalance in Pre-Vaccination State

Abstract Pre-vaccination or baseline antibody levels need not to be zero. Examples of infectious diseases for which this can be the case are tetanus, diphtheria, pertussis and tick-borne encephalitis. Imbalance in pre-vaccination state, i.e. a difference in baseline antibody levels between vaccine groups, can complicate the interpretation of a difference in post-vaccination antibody values. A standard approach to this problem is analysis of covariance. But in case of antibody values, one of the assumptions underlying this analysis, homoscedasticity, is not met. The larger the baseline value, the smaller the standard deviation of the error term. In this chapter, a solution to this problem is offered. It is shown that the heteroscedasticity can be modelled. A variance model is derived, and it is demonstrated how this model can be fitted with SAS.

5.1 Imbalance in Pre-Vaccination State

For some infectious diseases, pre-vaccination antibody levels are not zero. Not all vaccines offer lifelong protection, and a number of diseases require re-vaccinations throughout life. Antibody levels prior to re-vaccination with, for example, a tetanus, a diphtheria, a pertussis or a tick-borne encephalitis vaccine will therefore often not be zero. If pre-vaccination (baseline) antibody levels are not zero, then the post-vaccination values do not only express the immune responses to the vaccination, but also the subjects' pre-vaccination state. This can complicate the interpretation of a difference in post-vaccination antibody values between vaccine groups. If there is imbalance in pre-vaccination state, i.e. if there is a difference in baseline antibody values between groups, then part of the post-vaccination difference can be explained by the pre-vaccination difference. In case of a positive imbalance in pre-vaccination state, the post-vaccination difference may overestimate the immunogenicity of an investigational vaccine.

A popular approach to correct for imbalance in pre-vaccination state is analysing the fold increases instead of the post-vaccination antibody values. In Sect. 3.3.3, it was argued that the reasoning behind this approach—if pre-treatment values are subtracted from post-treatment values, any bias due to baseline imbalance is

J. Nauta, *Statistics in Clinical and Observational Vaccine Studies*,
Springer Series in Pharmaceutical Statistics,
https://doi.org/10.1007/978-3-030-37693-2_5

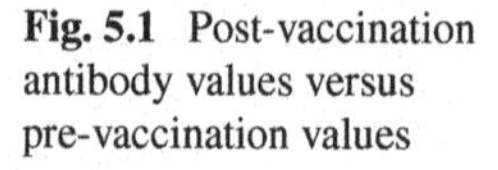
Fig. 5.1 Post-vaccination antibody values versus pre-vaccination values

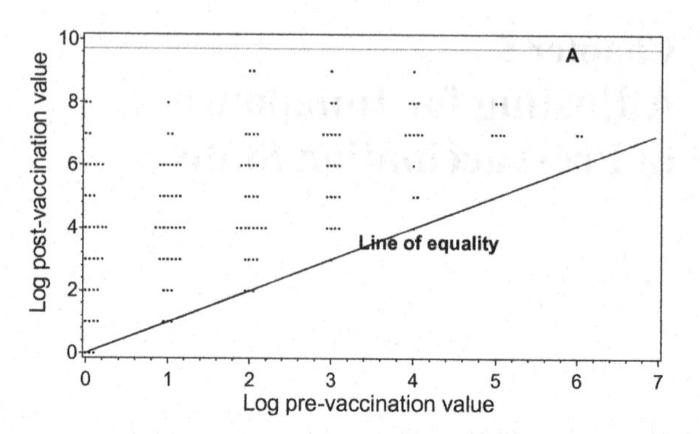

eliminated—is incorrect. If post- and pre-vaccination antibody values are positively correlated, and they usually are, then a positive baseline difference is predictive of a positive post-vaccination difference, but it is also predictive of a negative difference in mean fold increase. Thus, an analysis of fold increases does not solve the problem of imbalance in pre-vaccination state.

Consider the scatter plot in Fig. 5.1. Shown are log-transformed post-vaccination antibody values (y-axis) versus log-transformed pre-vaccination values (x-axis) for a given infectious disease. All points fall above or on the diagonal line of equality, because as a rule a post-vaccination antibody value will be larger than or equal to the pre-vaccination value. Because biologically there is a maximum to the number of antibodies that the body can produce, all points fall below a horizontal line, the asymptote A. The vertical distance between a point and the diagonal line is the log-transformed fold increase. The larger the log-transformed pre-vaccination antibody value, the smaller, on average, the log-transformed fold increase. Figure 5.1 is thus a graphical illustration of the negative correlation between pre-vaccination antibody values and fold increases.

Before proceeding to discuss two statistical techniques to adjust for baseline imbalance, the question needs to be addressed if such an adjustment is required. The answer is no. Statistical theory does not require baseline balance. In a randomized trial, baseline balance is not a peremptory requirement to yield valid results. Even with randomization, treatment groups will never be fully balanced with respect to all prognostic factors. If randomization is applied, groups will be equal on average, i.e. over all possible randomizations. Random between-group outcome differences are allowed for in the statistics calculate from the data, and the inference drawn from them is correct, on average.

Nevertheless, there is a general consensus that the credibility of a statistical analysis is increased if baseline imbalance for known prognostic factors is corrected for, as an improvement over just relying on the randomization.

5.2 Adjusting for Baseline Imbalance

There are two major statistical techniques to adjust for baseline imbalance: stratification and analysis of covariance.

In a stratified trial, subjects with a similar baseline value are assigned to the same stratum. If subjects are randomized per stratum, then treatment groups will be comparable with respect to the variable used for stratification. In clinical vaccine trials, however, pre-randomization stratification by baseline antibody value is rarely applied. One reason for this may be that this approach requires two instead of one baseline visit. During the first visit, a blood sample for baseline antibody titration is drawn. Baseline samples are then sent to a laboratory for antibody determination, which may take several weeks. And then, when the assay results have been received by the site, the subjects must return for a second baseline visit. If the vaccine contains different serotypes of the same organism, then pre-randomization stratification is also very difficult. An alternative to pre-randomization stratification is post-randomization stratification, with the strata being defined after the trial has been completed, during the statistical analysis of the data. A much applied approach is to divide the baseline values into the four quartiles, which then serve as strata. In both cases, pre- or post-vaccination stratification, a stratified analysis is performed, and the baseline imbalance is eliminated.

The second technique to adjust for baseline imbalance is analysis of covariance, sometimes referred to as regression control. (Analysis of covariance can be viewed as a limiting case of post-randomization stratification, with the observed baseline values as the strata.) Two parallel lines are fitted to the outcome data, one for the investigational treatment group and one for the control group, with regression on the baseline data. The treatment effect is then estimated by the vertical distance between the fitted lines. If there is baseline imbalance, a comparison of the outcome means will be affected by the difference in mean baseline value. Affected in the sense that the difference of the outcome means will overestimate the treatment effect. A comparison between the two groups for subjects with the same baseline value would be the solution. This is what analysis of covariance does.

5.3 Analysis of Covariance for Antibody Values

The simplest case of analysis of covariance for antibody values is the analysis of data of a study with only one vaccine. A linear regression model is fitted to the data with the log-transformed post-vaccination antibody value (y) as the response variable and the log-transformed pre-vaccination antibody value (x) as the predictor variable:

$$\mathbf{Y} = \beta_0 + \beta_1 X + \mathbf{E}_x.$$

The intercept β_0 and the slope β_1 are regression parameters, to be estimated from the data. If the observations are antibody titres and the standard log transformation is used, then the intercept β_0 is the mean of the distribution underlying the y's of the seronegative subjects, i.e. the subjects with a pre-vaccination antibody titre of D/2, with D the starting dilution factor. Then the slope β_1 is the average increase in the y's per dilution step, and

$$(\mathrm{D}/2)2^{\beta_0+\beta_1 \mathrm{x}}$$

is the geometric mean of the distribution underlying the post-vaccination antibody values of subjects with a log-transformed pre-vaccination value of x. Thus, the estimated regression parameters can be used for inference about the untransformed antibody values.

5.3.1 A Solution to the Problem of Heteroscedasticity

The residual $\mathbf{E}_x$ is the difference between the actual y and the value predicted by the fitted model. The usual assumption is that $\mathbf{E}_x$ is normally distributed about a mean value of 0 with variance σ_x^2. Another usual assumption is that of homoscedasticity. This is the assumption that σ_x^2 does not depend on x, but that $\sigma_x^2 = \sigma^2$. However, Fig. 5.1 shows that in case of log-transformed antibody values this assumption may not hold, but that σ_x^2 decreases with increasing x. A dependency of σ_x^2 on x is called heteroscedasticity. If this heteroscedasticity is ignored, i.e. if the regression model is fitted under the assumption of homoscedasticity, then the resulting parameter estimates, confidence intervals and P-values will be invalid. Fortunately, the heteroscedasticity can be modelled.

Let A be the upper limit for the log-transformed post-vaccination antibody values, i.e. the horizontal asymptote in Fig. 5.1. For normal data, an approximate formula for the range of the values to be observed is

$$\text{range} \approx \; c \times \text{standard deviation},$$

with the constant c often being set to either 4 or 6. The range for the y's for subjects with a log-transformed pre-vaccination value of x is (A-x), and thus

$$(\mathrm{A} - \mathrm{x}) \approx c\sigma_{\mathrm{x}},$$

or, after taking squares on both sides of the equation

$$\mathrm{A}^2 - 2\mathrm{A}\mathrm{x} + \mathrm{x}^2 \approx \mathrm{c}^2\sigma_{\mathrm{x}}^2.$$

This can be rewritten as

$$\sigma_x^2 \approx \sigma^2(1 + c_1 x + c_2 x^2), \tag{5.1}$$

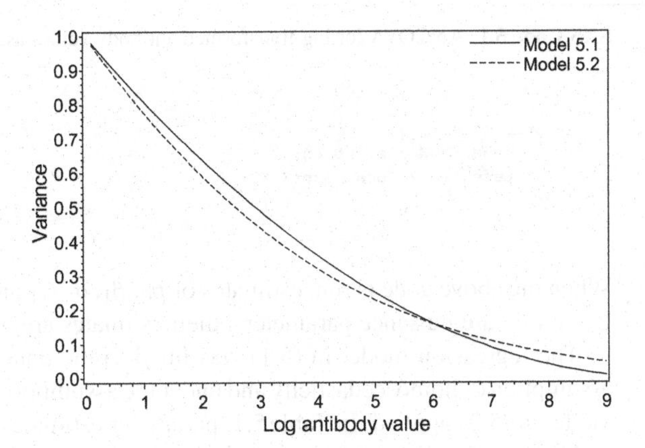

Fig. 5.2 Graphical comparison of variance models (5.1) and (5.2)

where

$$\sigma^2 = (\mathrm{A/c})^2, \quad \mathrm{c}_1 = -2/\mathrm{A}, \quad \mathrm{c}_2 = 1/\mathrm{A}^2.$$

Equation (5.1) thus gives a model for the variance of $\mathbf{E}_x$.

5.3.2 *Fitting the Variance Model for Heteroscedasticity*

The variance model in (5.1) cannot be fitted with SAS, but a similar model can. This model is

$$\sigma_x^2 = \sigma^2 \exp(C_1 x + C_2 x^2). \tag{5.2}$$

Consider the case that

$$A = 10, \quad c_1 = -2/10 = -0.2, \quad c_2 = 1/10^2 = 0.01.$$

This model is plotted in Fig. 5.2, where for convenience σ has been set to 1. Also plotted in Fig. 5.2 is the model in (5.2) with

$$C_1 = -0.24, \quad C_2 = -0.011.$$

The two curves are almost coincident.

The model in (5.2) has the nice property that it cannot be negative or zero, which is a guarantee for more stable variance estimates. Below a SAS code to fit this variance model is given.

SAS Code 5.1 ANCOVA for log-transformed antibody values assuming heteroscedasticity

```
proc mixed;
   model Y=X / solution;
   repeated / local=exp(X X2);
run;
```

When the above code is run, estimates of β_0, β_1, σ, C_1 and C_2 are returned. Because C_1 and C_2 are nuisance parameters, their estimates are of no special interest.

The regression model in (5.1) was fitted to the data of Fig. 5.1, both under the assumption of homoscedasticity and under the assumption that for the σ_x^2 the variance model in (5.2) applied. In Table 5.1, parameter estimates are given.

Not surprisingly, the estimate of σ under the assumption of homoscedasticity is smaller than the estimate under the assumption of heteroscedasticity.

If the standard log transformation is used, the intercept β_0 is the mean of the distribution underlying the y's of the seronegative subjects. This parameter is often of special interest because the seronegative subjects are the ones most in need of improved immunity. Alternative estimates of β_0 and σ are therefore the sample mean and standard deviation of the y's of the seronegative subjects, which are also given in Table 5.1. That this latter standard deviation is an estimate not only of σ_0 but also of σ is easy to see, because

$$\sigma_0^2 = \sigma^2 \exp(C_1 0 + C_2 0^2) = \sigma^2.$$

The estimator for σ_0 based on the regression approach is the more precise of the two. This can be seen by comparing the two standard errors. The standard error based on the y's of the seronegative subjects is

$$2.212/\sqrt{55} = 0.298.$$

In contrast, the standard error of the regression estimate of σ_0—taken from the SAS output (not shown)—is 0.218. The explanation for this difference is that the first standard error is based on only 55 observations, while the second one is based on all 165 observations.

5.3.3 ANCOVA for Comparative Clinical Vaccine Trials

Assume that in a randomized clinical vaccine trial the pre-vaccination geometric mean titres are 13.2 for the investigational vaccine group and 7.9 for the control vaccine group, expressing a moderate baseline imbalance in favour of the investigational vaccine.

Table 5.1 Regression parameter estimates for the data in Fig. 5.1

	n	β_0	β_1	σ
Assuming homoscedasticity	163	3.715	0.669	1.867
Assuming heteroscedasticity	163	3.765	0.642	2.151
Seronegative subjects only	55	3.818		2.212

The post-vaccination geometric mean titres are 286.3 and 112.8, respectively. The uncorrected post-vaccination geometric mean ratio thus is

$$\text{uncorrected } GMR = 286.3/112.8 = 2.54.$$

To correct for this baseline imbalance, the following regression model was fitted to the data:

$$\mathbf{Y} = \beta_0 + \beta_1 Group + \beta_2 X + \mathbf{E}_x, \tag{5.3}$$

with $Group = 1$ if the subject was vaccinated with the investigational vaccine and $Group = 0$ if the subject was vaccinated with the control vaccine. Because parallel regression lines are assumed, β_1 is the expected difference between the two groups for subjects with the same baseline titre. The baseline-corrected geometric mean ratio is

$$\text{baseline-corrected } GMR = 2^{\beta_1}.$$

The baseline-corrected geometric mean ratio should be smaller than the uncorrected one.

The model in (5.3) was fitted under the assumption of heteroscedasticity, using the following SAS code:

SAS Code 5.2 ANCOVA for log-transformed antibody values assuming parallel regression lines

```
proc mixed;
   model Y=Group X / solution;
   repeated / local=exp(X X2);
run;
```

SAS Output 5.2

Effect	Estimate	Std. Error	DF	t Value	Pr >|t|
intercept	4.1538	0.2148	243	19.34	<.0001
Group	0.7931	0.2769	243	2.86	0.0045
X	0.6079	0.07202	243	8.44	<.0001

The estimated baseline-corrected geometric mean ratio is

$$2^{0.7931} = 1.73,$$

a value, which is in indeed smaller than the uncorrected ratio, 2.54.

Above the assumption of parallel regression lines was made. If this assumption is in doubt, it can be checked by adding to the model the term for the interaction between *Group* and *X*.

SAS Code 5.3A ANCOVA for log-transformed antibody values, non-parallel regression lines

```
proc mixed;
   model Y=Group X Group*X / solution;
   repeated / local=exp(X X2);
run;
```

SAS Output 5.3A

Effect	Estimate	Std. Error	DF	t Value	Pr >|t|
intercept	3.9608	0.2326	242	17.03	<.0001
Group	1.1792	0.3306	242	3.57	0.0004
X	0.8217	0.1214	242	6.77	<.0001
Group*X	-0.3157	0.1473	242	-2.14	0.0331

The interaction term is significant. For subjects vaccinated with the investigational vaccine, the regression line is somewhat flatter than that for subjects vaccinated with the control vaccine.

When lines are not parallel, the distance between the lines, and thus the baseline-corrected geometric mean ratio, becomes dependent on x. When the interaction model is fitted to the data, β_1 is the expected difference between the two groups for seronegative subjects. For the example data, the estimate for β_1 and thus that for the geometric mean ratio

$$2^{1.1792} = 2.26$$

is statistically significant. It may, however, be that the group of special interest is not the seronegative subjects but, say, the subjects with as baseline value the threshold of protection T_P. In that case, the regression model to be fitted becomes

$$\mathbf{Y} = \beta_0 + \beta_1 Group + \beta_2 Z + \beta_3 Group \times Z + \mathbf{E}_x, \quad (5.4)$$

where $z = (x - \log T_P)$. This model can be fitted with the following SAS code:

SAS Code 5.3B ANCOVA for log-transformed antibody values, non-parallel regression lines

```
proc mixed;
   model Y=Group Z Group*Z / solution;
   repeated / local=exp(X X2);
run;
```

Note that here also the variance of the error term is defined as a function of X, not of Z.

SAS Output 5.3B

Effect	Estimate	Std. Error	DF	t Value	Pr >|t|
intercept	6.4260	0.3219	242	19.96	<.0001
Group	0.2320	0.3761	242	0.62	0.5379
Z	0.8217	0.1214	242	6.77	<.0001
Group*Z	-0.3157	0.1473	242	-2.14	0.0331

The only two estimates that change are those for β_0 and β_1. For subjects with the baseline value equal to the threshold of protection, the geometric mean ratio

$$2^{0.2320} = 1.17$$

is not statistically significant.

Chapter 6
Vaccine Equivalence and Non-inferiority Trials

Abstract Many vaccine immunogenicity trials are conducted in an equivalence or non-inferiority framework. The objective of such trials is to demonstrate that the immunogenicity of an investigational vaccine is comparable or not less than that of a control vaccine. Four types of vaccine equivalence and non-inferiority designs are distinguished: vaccine bridging trials, combination vaccine trials, vaccine concomitant use trials and vaccine lot consistency trials. In this chapter, the statistical analysis of the data of such trials is covered, both for trials with a geometric mean response as endpoint and trials with seroprotection or seroconversion as endpoint. The standard analysis of lot consistency data is known to be conservative, but a simple formula is presented which can be used to decide if the lot sample sizes guarantee that the actual type I error rate of the trial is sufficiently close to the nominal level. The chapter is concluded with a discussion of sample size estimation for vaccine equivalence and non-inferiority trials, including lot consistency trials.

6.1 Equivalence and Non-inferiority

So far, it has been silently assumed that the objective of the trial was to demonstrate that one vaccine is superior to another with respect to the induced immunogenicity. Trials with this objective are called superiority trials. In an equivalence trial, the objective is not to demonstrate that two vaccines are different but that they are more or less equally immunogenic, while in a non-inferiority trial the objective is to demonstrate that one vaccine is not less immunogenic than the other.

Wang et al. identify four types of equivalence and non-inferiority designs for vaccine immunogenicity trials: vaccine bridging trials, combination vaccine trials, vaccine concomitant use trials and vaccine lot consistency trials [25]. In a *vaccine bridging trial*, two formulations of a vaccine are being compared. The reason for the study could be a change in the manufacturing process, a change in the vaccine formulation, or a change in storage conditions of the vaccine. An example of the first is when a second manufacturing site is opened, and it has to be demonstrated that the immunogenicity of the formulation produced at the new site is comparable to that of the formulation produced at the old site. An example of the second reason is the

J. Nauta, *Statistics in Clinical and Observational Vaccine Studies*,
Springer Series in Pharmaceutical Statistics,
https://doi.org/10.1007/978-3-030-37693-2_6

removal of a constituent from the formulation. Until recently many vaccines contained thiomersal as preservative. After the thiomersal controversy (see Sect. 12.1), manufacturers were requested to remove the preservative from their childhood vaccines and today all childhood vaccines are thiomersal-free.

In a *combination vaccine trial*, immune responses are compared between a combined vaccine and the separate but simultaneously administered monovalent vaccines. A combination vaccine is usually intended to reduce the number of injections required. In 2005, the Food and Drug Administration (FDA) approved the licensure of a combination MMRV (measles, mumps, rubella, varicella) vaccine for children aged twelve months through twelve years, as alternative for two separate MMR and V vaccines. (However, soon after licensure the manufacturer of the combination vaccine withdrew the vaccine because of a safety issue, see Sect. 12.1.) Before a combination vaccine is licensed it has to be demonstrated that the combination is not less immunogenic or less safe than the monovalent vaccine. In the past, combination of whole live vaccines resulted in a reduced immune response due to immunological inference between vaccine viruses.

A *vaccine concomitant use trial* is used to compare the concomitant administration of two or more vaccines and the separate administration of the vaccines. The intention is usually to see if the number of vaccination visits can be reduced.

A vaccine manufacturer who wants to license his vaccine must demonstrate that the manufacturing process is stable, i.e. that consistent lots can be produced. This has to be demonstrated by both analytical and clinical testing. The clinical testing is done in a so-called *vaccine lot consistency trial*, and the objective of such a study is to show that the lots are similar with respect to the induced immunogenicity.

The first three types of trials are usually designed as non-inferiority studies, while vaccine lot consistency trials are an example of an equivalence study. But, because the concept equivalence was introduced before that of non-inferiority, first equivalence studies are discussed.

6.2 Equivalence and Non-inferiority Testing

6.2.1 Basic Concepts

Let $\Delta = \mu_1 - \mu_0$ be the difference between (the expected means of) two treatments. In a superiority trial, the null hypothesis is $H_0 : \Delta = 0$, which is usually tested against the two-sided alternative $H_1 : \Delta \neq 0$, or, less common, against a one-sided alternative, say, $H_1 : \Delta > 0$. In contrast, in an equivalence trial, the objective is not to demonstrate that two treatments are different but that they are similar, meaning that $\Delta \approx 0$. Similar, because to proof exact equality would be impossible. To be more precise, the objective of an equivalence trial is to demonstrate that, on average, two treatments differ no more than by a fixed amount M, the equivalence margin. The two treatments are considered equivalent if $|\Delta| < M$. To demonstrate equivalence of two treatments Schuirmann's two one-sided tests (TOST) procedure is often used [26]. There are two null hypotheses associated with the procedure that the difference between the two treatments exceeds the equivalence margin:

$$H_{01} : \Delta \leq -M \quad \text{and} \quad H_{02} : \Delta \geq M.$$

These null hypotheses are tested against the alternatives

$$H_{11} : \Delta > -M \quad \text{and} \quad H_{12} : \Delta < M.$$

To demonstrate equivalence both null hypotheses have to be rejected. If both H_{01} and H_{02} are tested at the significance level α, this approach corresponds to checking that the two-sided $100(1-2\alpha)\%$ confidence interval for Δ lies within the equivalence range $-M$ to $+M$.

A somewhat controversial issue in equivalence trials is the choice of the significance level α, with the question being whether the level should be set to 0.05 or 0.025. Schuirmann himself suggested $\alpha = 0.05$, and in pharmacokinetics studies, for example, this has become the standard. In non-phase I clinical studies, however, the preference seems to be $\alpha = 0.025$. In a superiority trial in which an active treatment is compared with a placebo, most regulatory agencies will require a two-sided significance level of 0.05. Given that the only outcome of interest is where the active treatment is significantly better than the placebo, the risk for the regulatory agency is at most 0.025. In an equivalence trial, the risk for the agency is that the true null hypothesis (either H_{01} or H_{02}) is falsely rejected. If both null hypotheses are tested at significance level, this risk is at most α. And for this reason, it is often argued that in equivalence trials the significance level should be set to 0.025, for the sake of consistency. To stress this, in this chapter equivalence (and non-inferiority) will be tested at the significance level $\alpha/2$.

Non-inferiority trials are a special case of equivalence trials, the one-sided version, so to speak. To objective of a non-inferiority trial is to demonstrate that one treatment is not less than another by more than a small amount, M, the non-inferiority margin. This is tested by the comparison: reference vaccine minus test vaccine. Let $\Delta = \mu_{\text{ref}} - \mu_{\text{test}}$. The null hypothesis of a non-inferiority trial is then that H_0: $\Delta \geq M$, which is tested against the alternative H_1: $\Delta < M$. If the null hypothesis is tested at the significance level $\alpha/2$, this procedure leads to the same conclusion as checking that the upper bound of the one-sided $100(1 - \alpha/2)\%$ confidence interval for Δ falls to the left of M.

As pointed out in ICH guideline E9, an equivalence or a non-inferiority trial must be analysed according to the per-protocol principle [27]. In superiority trials, the primary analysis is usually according to the intention-to-treat principle, because it tends to avoid over-optimistic estimates of efficacy resulting from a per-protocol analysis (non-compliers in the intention-to-treat will generally diminish the estimated treatment effect). In an equivalence or non-inferiority trial, an intention-to-treat analysis is generally not conservative, but the per-protocol analysis is.

6.2.2 Normal Data

An alternative way to explain testing for equivalence is that the null hypothesis $H_0 : |\Delta| \geq M$ is tested against the alternative $H_1 : |\Delta| < M$. If both groups of n_1 and n_0 observations are normally distributed, this null hypothesis can be tested using the following test statistic:

$$Z_{EQ} = \frac{M - |\hat{\Delta}|}{SE(\hat{\Delta})}, \tag{6.1}$$

where $\hat{\Delta} = x_{1.} - x_{0.}$, the difference between the arithmetic means of the two groups of observations, and $SE(\hat{\Delta})$ the standard error of this difference:

$$SE_P(\hat{\Delta}) = \sqrt{SD_1^2/n_1 + SD_0^2/n_0},$$

where SD_1 and SD_0 are the sample standard deviations of the two groups. If $|\hat{\Delta}| > M$, the data are in favour of the null hypothesis, in which case Z_{EQ} will be negative. If, on the other hand, $|\hat{\Delta}| < M$ then Z_{EQ} will be positive, and the smaller $|\hat{\Delta}|$ the larger Z_{EQ}. The null hypothesis is thus rejected for large positive values of the test statistic, which can be compared with the $100(1 - \alpha/2)$th percentile of the standard normal distribution. Note that Z_{EQ} is by definition a one-sided test statistic. If equal variances are assumed, an alternative estimator for the standard error is

$$SE(\hat{\Delta}) = SD\sqrt{1/n_1 + /n_0}, \tag{6.2}$$

where SD is the pooled sample standard deviation. If this standard error is used, the value for the test statistic can be compared with the $100(1-\alpha/2)$th percentile of the t-distribution with $(n_1 + n_0 - 2)$ degrees of freedom.

In a non-inferiority trial, the null hypothesis H_0: $\Delta \geq M$ is tested against the alternative H_1: $\Delta < M$. This null hypothesis can be tested using the following test statistic:

$$Z_{NI} = \frac{M - \hat{\Delta}}{SE(\hat{\Delta})}. \tag{6.3}$$

The data are in favour of the null hypothesis if $\hat{\Delta} \geq M$, i.e. if $M - \hat{\Delta} \leq 0$, in which case Z_{NI} will be negative. Z_{NI} will be positive if the data are in favour of the alternative hypothesis, i.e. if $\hat{\Delta} < M$. Thus here also, the null hypothesis is rejected for large positive values of the test statistic. Values for Z_{NI} can be compared either with $100(1 - \alpha/2)$th percentile of the standard normal distribution, or the $100(1-\alpha/2)$th percentile of the t-distribution with $(n_1 + n_0 - 2)$ degrees of freedom when the standard error (6.2) is used.

6.2.3 The Confidence Interval Approach

If $H_0 : |\Delta| \geq M$ is tested at the significance level $\alpha/2$, then the null hypothesis that the two treatments are not equivalent is rejected if $Z_{EQ} > z_{1-\alpha/2}$. This will be the case if and only if

$$-M < \hat{\Delta} \pm z_{1-\alpha/2} SE(\hat{\Delta}) < M.$$

Thus, testing for equivalence involves checking that the two-sided $100(1-\alpha)\%$ confidence interval for Δ falls in the equivalence range $-M$ to $+M$. It is this confidence interval approach that is most often used in scientific publications rather than the hypothesis testing approach. If equal variances are assumed then the two-sided $100(1-\alpha)\%$ confidence interval can be based on the t-distribution, in which case it should be checked that

$$-M < \hat{\Delta} \pm t_{1-\alpha/2;n_1+n_0-2} SE_P(\hat{\Delta}) < M.$$

The null hypothesis that one treatment is less than another can be tested by checking that the upper bound of the one-sided $100(1-\alpha/2)\%$ confidence interval for Δ falls to the left of M:

$$\hat{\Delta} + z_{1-\alpha/2} SE(\hat{\Delta}) < M.$$

6.3 Geometric Mean Response as Outcome

When the outcome measure of a clinical vaccine trial is a geometric mean response—a geometric mean titre or geometric mean concentration—then the parameter of interest is the geometric mean ratio θ, the ratio of the geometric means e^{μ_1} and e^{μ_0} of the distributions underlying the immunogenicity values. Requiring that the confidence interval for the ratio θ falls within a pre-specified range is the same as requiring that the confidence interval for the difference Δ of the arithmetic means μ_1 and μ_0 of the distributions underlying the log-transformed values falls within the log-transformed range. This means that equivalence can be stated both in terms of θ and Δ: if M_L to M_U is an equivalence range for the ratio θ, then $\log M_L$ to $\log M_U$ is an equivalence range for the difference Δ, and vice versa. Because it is a matter of indifference which of the two vaccines is called 'vaccine A' and which is called 'vaccine B', the parameter of interest can be both θ and $1/\theta$. For this reason, an equivalence range for θ is usually defined as $1/M$ to M. This corresponds to setting the equivalence range for Δ to the symmetrical range $-\log M$ to $\log M$.

Example 6.1 Joines et al. report the results of a combination vaccine trial [28]. They compared a combination hepatitis A and B vaccine with the monovalent vaccines. Both infectious diseases can be fatal. The major cause of hepatitis A is ingestion of faecally contaminated food or water. Hepatitis B is a sexually transmitted disease. In the trial, 829 adults were randomized to receive either the combination vaccine by three separate intramuscular injections in the deltoid or two separate intramuscular injections in the deltoid with hepatitis A and three separate injections in the deltoid of the opposite arm with hepatitis B vaccine. The primary analysis was a non-inferiority analysis for the incidences of severe soreness, a safety endpoint. The secondary analysis was an equivalence analysis for the seroconversion rates for hepatitis A and the seroprotection rates for hepatitis B. In this example, the focus will be on the secondary analysis, the analysis of the month 7 immunogenicity data. Antibody titres to hepatitis A (anti-HAV) were determined using an enzyme immunoassay kit, and seroconversion for hepatitis A was defined as an anti-HAV titre $\geq$33 mIU/ml. Antibody titres to hepatitis B (anti-HBs) were determined using a radioimmunoassay kit, and seroprotection for hepatitis B was defined as an anti-HBS titre $\geq$10 mIU/ml. In total, 533 subjects were included in the per-protocol sample, 264 vaccinated with the combination vaccine and 269 vaccinated with monovalent vaccines. The main reason for exclusion from the per-protocol sample was being seropositive to hepatitis A or hepatitis B at baseline.

At month 7, the anti-HBs geometric mean titre and geometric standard deviation were 2,099 and 6.8 in the combination vaccine group and 1,871 and 9.5 in the monovalent vaccines group. Hence, the geometric mean ratio was

$$GMR = 2{,}099/1{,}871 = 1.12.$$

Because the geometric mean titre was not a secondary outcome, no equivalence range for the ratio θ was specified. For illustrative purposes, here, the range 0.67 to 1.5 will be used. The arithmetic means of the log-transformed antibody titres were

$$\log 2{,}099 = 7.65 \quad \text{and} \quad \log 1{,}871 = 7.53.$$

The corresponding standard deviations were

$$\log 6.8 = 1.92 \quad \text{and} \quad \log 9.5 = 2.25,$$

giving as value for the pooled standard deviation 2.09. Standard error (6.2) takes the value

$$2.09\sqrt{1/264 + 1/269} = 0.18.$$

Thus, the value for the test statistic (6.1) is

$$Z_{EQ} = \frac{0.41 - |7.65 - 7.53|}{0.18} = 1.61,$$

where $0.41 = \log 1.5$. The corresponding P-value, from the t-distribution with $(264 + 269 - 2) = 531$ degrees of freedom, is 0.054. Thus, the null hypothesis that the combination vaccine and the monovalent vaccine are not equivalent with respect to the induced anti-HBs responses cannot be rejected.

The 97.5th percentile of the t-distribution with 531 degrees of freedom is 1.964. Hence, the lower and upper bounds of the two-sided 95% confidence interval for Δ are

$$(7.65 - 7.53) - 1.964(0.18) = -0.23$$

and

$$(7.65 - 7.53) + 1.964(0.18) = 0.47.$$

By taking the antilogs of these limits, the limits of the 95% confidence interval for the geometric mean ratio θ are obtained. The resulting interval (0.79, 1.60) does not fall in the equivalence range 0.67 to 1.5.

Assume that another secondary objective was to demonstrate that the combination vaccine is non-inferior to the monovalent vaccine with respect to the induced anti-HAV responses. The month 7 anti-HAV geometric mean titre and geometric standard deviation were 4,756 and 3.1 for the combination vaccine group and 2,948 and 2.5 in the monovalent vaccines group. Hence, $GMR = 0.62$. The arithmetic means of the log-transformed antibody titres were

$$\log 4{,}756 = 8.47 \quad \text{and} \quad \log 2{,}948 = 7.99$$

with standard deviations

$$\log 3.1 = 1.13 \quad \text{and} \quad \log 2.5 = 0.92,$$

respectively. The pooled standard deviation is 1.03, with $SE(\hat{\Delta}) = 0.09$. The lower bound of the one-sided 97.5% confidence interval for Δ is

$$(7.99 - 8.47) + 1.964(0.09) = -0.30,$$

which corresponds to an upper bound of

$$e^{-0.30} = 0.74$$

for the geometric mean ratio θ. Because the upper bound falls to the left + of the non-inferiority bound 1.5, the null hypothesis that the combination vaccine is inferior to the monovalent vaccine can be rejected.

6.4 Seroresponse Rate as Outcome

Equivalence or non-inferiority of two vaccines with a seroprotection or a seroconversion rate (i.e. a proportion) as endpoint is usually demonstrated by means of the confidence interval approach, with the risk difference as the parameter of interest. In case of non-small group sizes, the asymptotic confidence interval based on the

Wilson method can be used (Sect. 3.5.2), while in case of small group sizes the exact interval based on Barnard's test (Sect. 3.5.3) should be used. Often used equivalence and non-inferiority margins for the risk difference are 0.05 (5%) and 0.10 (10%).

Example 6.1 (continued) The equivalence of the combination vaccine and the monovalent was to be demonstrated using the seroconversion rates for hepatitis A and the seroprotection rates for hepatitis B. For hepatitis A, equivalence was to be concluded if the two-sided 95% confidence interval for the seroconversion risk difference would fall in the equivalence range −0.05 to 0.05. The observed seroconversion rates were

$$267/269 = 0.993 \quad \text{and} \quad 263/264 = 0.996$$

for the combination vaccine group and the monovalent vaccines group, respectively. The estimated risk difference thus was

$$0.993 - 0.996 = -0.003.$$

The 95% Wilson-type confidence interval for the risk difference,

$$(-0.023, 0.014)$$

is contained in the predefined equivalence range, and equivalence can be concluded. With SAS, it is possible to test the two one-sided hypotheses associated with proving equivalence using the Wilson-type test statistic on the left-hand side of Eq. (3.7). The SAS code is

SAS Code 6.1 TOST Procedure for rates using the Wilson-type test statistic

```
proc freq;
   table Vaccine*Seroconverted /
      riskdiff (equivalence method=fm margin=0.05) alpha=0.025;
run;
```

The TOST P-values for the example above are 0.0004 and <0.0001. Both null hypotheses can be rejected. Note that the code can also be used to find the Wilson-type confidence limits by trial and error. The lower confidence limit is the smallest value for the lower margin for which the corresponding TOST P-value is ≥ 0.025. For MARGIN = 0.022 the P-value is 0.0306, for MARGIN = 0.023 the P-value is 0.0262, and for MARGIN = 0.024 the P-value is 0.0224. From this, it can be concluded that the lower limit of the Wilson-type confidence limit is −0.023.

6.5 Vaccine Lot Consistency Trials

Both FDA/CBER and the European Medicines Agency (EMA) require, prior to licensure of a vaccine, proof that the vaccine production process is stable and that consistent lots can be produced. As part of this requirement, a clinical study must be performed, a so-called lot consistency trial. The objective of a vaccine lot consistency trial is to show that the, preferably consecutively produced, lots (batches) are similar with respect to the induced immunogenicity. Subjects are randomly assigned to be vaccinated with vaccine from one of three lots. The post-vaccination blood samples of the subjects are assayed, and the antibody values are compared between the three lots. Lot consistency is concluded if all three pair-wise post-vaccination geometric mean ratios are close to one. Vaccine lot consistency trials are thus an example of an equivalence study.

6.5.1 *The Confidence Interval Method*

The most frequently applied method to demonstrate lot consistency is to calculate two-sided $100(1-\alpha)\%$ confidence intervals for the three pair-wise geometric mean ratios. If all three confidence intervals fall within the predefined equivalence range, lot consistency is concluded.

Example 6.2 Nauta discusses the statistical analysis of the data of influenza vaccine lot consistency trials [29]. The example he uses is a lot consistency trial with a virosomal influenza vaccine. (Virosomes are haemagglutinin and neuraminidase antigens linked to globular lipid membranes (liposomes), which are believed to have an adjuvant effect.) Prior to unblinding of the database, the Blind Review Committee excluded 10 (2.7%) of the 373 randomized subjects from the per-protocol sample. Anti-HA antibody titres were determined using the haemagglutination inhibition (HI) test. As log transformation, the standard log transformation with $D = 10$ was used. The equivalence range for the pair-wise geometric mean ratios was predefined to be 0.35 to 2.83. In Table 6.1, the results for the A-H3N2 strain are summarized. For lot #2 versus lot #1, the pooled standard deviation is 1.585. The 97.5th percentile of the t-distribution with $(123+123-2) = 244$ degrees of freedom is 1.970. Hence, the lower and the upper bound of the two-sided 95% confidence interval for Δ are

$$(5.02 - 5.27) - 1.970\sqrt{1.585(2/123)} = -0.648$$

and

$$(5.02 - 5.27) + 1.970\sqrt{1.585(2/123)} = 0.148.$$

By taking the antilogs (to the base 2), the two-sided 95% confidence intervals for the geometric mean ratio for lot #2 versus lot #1 is obtained:

Table 6.1 Summary statistics of an influenza vaccine lot consistency trial (A-H3N2 strain) [29]

	Lot #1 ($n = 123$)	Lot #2 ($n = 123$)	Lot #3 ($n = 117$)
Geometric mean titre	192.9	162.2	202.5
Arithmetic mean*	5.27	5.02	5.34
Standard deviation*	1.57	1.60	1.57

*: log-transformed antibody titres

$$(0.638, 1.108).$$

The two-sided 95% confidence intervals the geometric mean ratios for lot #3 versus lot #1 and for lot #3 versus lot #2 are

$$(0.796, 1.384) \quad \text{and} \quad (0.944, 1.651).$$

All three confidence intervals fall in the predefined equivalence range 0.38 to 2.83, and for the A-H3N2 strain lot consistency can be concluded.

6.5.2 *The Wiens–Iglewicz Test*

Proving lot consistency can also be formulated as a hypothesis testing problem. Wiens and Iglewicz developed a statistical test to demonstrate the equivalence of three treatments [30]. The Wiens–Iglewicz test, which requires normal data, can be used to demonstrate lot consistency. Let Δ_{ij} denote the difference between the expected means of the log-transformed antibody values of the ith and jth lot. To demonstrate lot consistency, the null hypothesis

$$H_0 : \max\{|\Delta_{ij}| \geq M\}$$

is tested against the alternative

$$H_1 : \max\{|\Delta_{ij}| < M\},$$

where M is the equivalence margin. Wiens and Iglewicz propose to test this null hypothesis by evaluating Z_{EQ} in (6.1) for all three pair-wise comparisons, and then to use the following min test statistic:

$$Z_{\min} = \min\left\{\frac{M - |\hat{\Delta}_{ij}|}{SE(\hat{\Delta}_{ij})}\right\},$$

where $\hat{\Delta}_{ij}$ is the difference between the arithmetic sampling means of the ith and the jth lot and $SE(\hat{\Delta}_{ij})$ the standard error of this difference. By definition, $Z_{\min}$ is a one-sided test, and the null hypothesis is rejected for large values for $Z_{\min}$. To test the above overall null hypothesis at the $\alpha/2$ significance level, the value of $Z_{\min}$ can be compared with the $100(1-\alpha/2)$th percentile of the standard normal distribution.

Example 6.2 (continued) For the A-H3N2 strain, the observed differences are

$$D_{12} = -0.25, \quad D_{13} = 0.07 \text{ and } D_{23} = 0.32.$$

The standard errors of these differences are

$$SE(D_{12}) = \sqrt{1.57^2/123 + 1.60^2/123} = 0.202$$
$$SE(D_{13}) = \sqrt{1.57^2/123 + 1.57^2/117} = 0.203$$
$$SE(D_{23}) = \sqrt{1.60^2/123 + 1.57^2/117} = 0.205.$$

This gives (with $M = \log_2 2.83 = 1.5$)

$$Z_{\min} = \min\left\{\frac{1.5-0.25}{0.202}, \frac{1.5-0.07}{0.203}, \frac{1.5-0.32}{0.205}\right\}.$$

The resulting value, 5.76, is highly statistically significant.

A nice property of the Wiens–Iglewicz test is that it is an overall test, and a correction for multiplicity is not needed. The drawback of the test is the complexity of its sampling distribution, which makes it difficult to evaluate the type I error rate. Because lots can be re-ordered, without loss of generality

$$0 \le \Delta_{12} \le \Delta_{13}.$$

Let

$$\Delta = \Delta_{13} \text{ and } \rho = \Delta_{13}/\Delta_{12},$$

with σ_i and n_i the within-lot standard deviation of the log-transformed antibody values and the sample sizes of the ith lot, and let

$$SE_{ij} = \sqrt{\sigma_i^2/n_i + \sigma_j^2/n_j}.$$

The type I error rate of $Z_{\min}$ depends on Δ/SE_{ij} and ρ. In Table 6.2, Monte Carlo simulation results are presented, for the simple case that

$$\sigma_1 = \sigma_2 = \sigma_3 \text{ and } n_1 = n_2 = n_3 = n.$$

The type I error rate is defined as

Table 6.2 Actual type I error rates lot consistency trial, for $\alpha = 0.05$ and critical value $z_{0.975}$

M/σ	n	ρ				
		0	0.25	0.5	0.75	1
0.25	500	0.011	0.029	0.044	0.033	0.011
	300	0.013	0.031	0.033	0.024	0.013
	100	0.001	0.001	0.002	0.002	0.002
0.5	500	0.011	0.050	0.048	0.047	0.010
	300	0.013	0.041	0.045	0.042	0.012
	100	0.010	0.032	0.042	0.033	0.015
1.0	500	0.013	0.048	0.049	0.053	0.014
	300	0.012	0.045	0.050	0.051	0.009
	100	0.015	0.044	0.048	0.044	0.015

$$\Pr(Z_{\min} > z_{1-\alpha/2} | \Delta = \sigma).$$

(The algorithm used for the simulation is similar to the algorithm for sample size estimation for lot consistency trials outlined in Sect. 6.7.3. The actual type I error rates are lowest for ρ close to 0.0 or 1.0 and highest for $\rho = 0.5$, and then they are close to the nominal error rate for non-small n and M/σ. Wiens and Iglewicz show that when

$$(M/\sigma)\sqrt{n/2} > 5.0$$

the actual type I error approaches the nominal one for $\rho = 0.5$. This is the same as requiring that

$$n > \frac{50}{(M/\sigma)^2}. \tag{6.4}$$

Equation (6.4) can be thus used to decide if the lot sample sizes guarantee that the actual type I error rate of the trial is sufficiently close to the nominal level. If $M/\sigma = 1.0$, then n must be greater than 50, if $M/\sigma = 0.5$, then n must be greater than 200 and when $M/\sigma = 0.25$, then n must be greater than 800. The simulation results in Table 6.2 are consistent with this.

Example 6.2 (continued) If it is assumed that $\sigma_1 = \sigma_2 = \sigma_3 = 1.6$, then $M/\sigma = 1.5/1.6 = 0.94$. To secure a non-conservative actual type I error rate, the group size per lot should be at least

$$\frac{50}{0.94^2} = 57,$$

which was the case.

The Wiens–Iglewicz approach and the confidence interval approach yield near similar results. The two approaches differ only in the standard errors being used and the critical value (derived from the standard normal distribution versus the t-distribution).

6.6 Discussion

Equivalence and non-inferiority trials are not uncontroversial. There are several reasons for this. One, perhaps the most important, reason is that it is often difficult to justify the choice of the margin. An overlay strict margin will require a prohibitively large sample size, while a too large margin will not be clinically meaningful. (That a margin leads to a too large sample size is, admittedly, not a very strong argument. Indeed, what matters is the clinical relevance of the differences that the margin allows.) Consider the problem of deciding an equivalence range for the geometric mean ratio. The range 0.67 to 1.5 is generally considered to be a reasonable one, not too wide but also not too strict. A proper justification for this range would take into account the strength of the relationship between the antibody measurements and the probability of clinical protection from infection. If the protection curve (see Sect. 11.3) would be a steep one, then a very small range would be appropriate, while a less steep one would allow a broader range. The difficulty is that in practice the relationship is seldom known with sufficient detail.

Many vaccines contain antigen of more than one serotype. In that case, equivalence or non-inferiority must usually be demonstrated for all serotypes. A much applied approach is to demonstrate equivalence or non-inferiority at the $\alpha/2$ significance level for each of the serotypes. The intersection-union (IU) principle then allows that equivalence or non-inferiority is claimed on vaccine level, and no multiplicity correction is needed. This approach is known to be conservative under many circumstances. Kong et al. propose *min* tests for vaccine equivalence and non-inferiority trials with multiple serotypes [31, 32]. These tests take the correlation between the endpoints into account. Simulation results, however, show that for trials with multiple binomial endpoints the min test leads to only modest increases in power.

A statistical novelty of recent date is simultaneous testing of non-inferiority and superiority [33]. Basically, if in a non-inferiority trial the null hypothesis is rejected, one can proceed to test the null hypothesis for superiority. In a superiority trial, if the null hypothesis is not rejected, one can proceed with a test for non-inferiority. The strategy can be justified by either the IU principle or the closed testing principle, and no multiplicity adjustment is needed [34, 35]. The strategy is not undisputed. One of its critics, Ng, argues that the strategy allows an investigational treatment to claim superiority by chance alone without risking the non-inferiority claims, which will increase the number of erroneous claims of superiority [36]. Despite this criticism, the strategy has quickly found its way into clinical development. Leroux-Roels et al. compare an intradermal (injected between the layers of the skin) trivalent inactivated split-virion influenza vaccine with an intramuscular control vaccine [37]. They conclude that the intradermal vaccine induces non-inferior humoral immune responses against all three virus strains included in the vaccines, because all three two-sided 95% confidence intervals for the geometric mean ratios fell above the pre-specified non-inferiority margin. In addition, they conclude superior responses against both A strains, because for these two strains the confidence interval fell above 1.0. This second conclusion is disputable. In the Statistical Methods section, the authors explain that non-inferiority was tested at vaccine level (i.e. to be demonstrated for *all* strains contained in the vaccines), but that superiority was tested per strain, at the two-sided significance level 0.05. As argued above, because non-inferiority was to be shown at vaccine level a multiplicity correction was not needed. But because superiority could have been claimed for any number of strains, here a multiplicity correction should have been applied. Thus, contrary to what was claimed, the statistical analysis does not warrant the conclusion of a superior response against both A strains.

6.7 Sample Size Estimation

6.7.1 Comparing Two Geometric Mean Responses

The statistical power of a clinical vaccine trial with the primary objective to demonstrate that the immunogenicity induced by an investigational vaccine is equivalent or non-inferior to that induced by a control vaccine, and with either a geometric mean titre or a geometric mean concentration as outcome is, apart from the sample size and the significance level, dependent upon:

1. The equivalence/non-inferiority margin M.
2. The within-group standard deviation σ of the log-transformed immunogenicity values.
3. The difference $\Delta = \mu_{\text{ref}} - \mu_{\text{test}}$, with μ_{ref} and μ_{test} the arithmetic means of the probability distributions underlying the log-transformed immunogenicity values of the reference and the test vaccine groups.

As will be shown below, the statistical power is profoundly sensitive to assumptions about Δ, which makes sample size estimation for vaccine equivalence and non-inferiority trials a challenging exercise. To power trials on the assumption of a zero difference between the vaccines is therefore not recommended. Better is to assume some amount variation between the vaccines.

Under the usual assumption that the log-transformed immunogenicity values are normally distributed, the statistical power of equivalence and non-inferiority trials should be estimated from non-central t-distributions. For equivalence trials sample size formula (5.5) in the book by Julious applies [18]. The formula can be solved with the PROC POWER.

Example 6.1 (continued) An investigator wishes to know the statistical power of the trial for the combination vaccine versus the monovalent hepatitis B vaccine. The equivalence margin for Δ is $M = \log 1.5 = 0.41$. Assume that for log-transformed anti-HBs titres the convention is to set σ to 2.0. For a first sample estimate Δ is set to 0.0. The desired statistical power is 0.90. The required sample size can be calculated with the following SAS Code:

SAS Code 6.2A Sample size estimation for an equivalence trial with normal data

```
proc power;
    twosamplemeans test=equiv_diff alpha=0.025
    meandiff=0  stddev=2.0
    lower=-0.41 upper=0.41
    power=0.9 npergroup=.;
run;
```

When this code is run, a required sample size of 620 subjects per group is found. Next, the investigator wishes to study the robustness of this estimate when it is assumed that the combination vaccine is somewhat less immunogenic than the monovalent vaccine, i.e. when it is assumed that difference Δ is less than zero or, which is the same, the geometric mean ratio θ is less than one. He proposes two values for the ratio, $\theta = 0.95$ and $\theta = 0.90$. These values correspond to $\Delta = 0.051$ and $\Delta = 0.105$. The robustness of the above sample size estimation can be inspected with the following SAS code:

SAS Code 6.2B Statistical power of an equivalence trial with normal data

```
proc power;
    twosamplemeans test=equiv_diff alpha=0.025
    meandiff=0, -0.051, -0.105 stddev=2.0
    lower=-0.41 upper=0.41
    npergroup=620 power=.;
run;
```

The SAS output is given below.

SAS Output 6.2B

```
            Computed Power
                Mean
     Index      Diff       Power
         1      0.000      0.900
         2     -0.051      0.866
         3     -0.105      0.760
```

If θ is assumed to be 0.9 instead of 1.0, the statistical power drops from 0.90 to 0.76. This demonstrates the critical dependency of the statistical power on Δ. If the value of 0.9 for θ would be considered to be more likely, then a sample size of 906 subjects per group would be required to be secured of a statistical power of 0.9.

The power of a non-inferiority trial with normal data is formula (6.6) in the book by Julious [18]. This formula can be evaluated with either SAS Code 6.2A or SAS Code 6.2B, with LOWER set to a very large negative value, say −999.

Example 6.1 (continued) Assume that the investigator also wishes to know the statistical power of the trial for combination vaccine versus the monovalent hepatitis A vaccine. The within-group standard deviation σ is set to 1.25, Δ to 0.0 and the non-inferiority margin to 0.41. First it is assumed that Δ equals zero, and the investigator would like to know the required numbers of subjects to be secured of a statistical power of 0.9.

SAS Code 6.3 Statistical power of a non-inferiority trial with normal data

```
proc power;
   twosamplemeans test=equiv_diff alpha=0.025
   meandiff=0 stddev=1.25
   lower=-999 upper=0.41
   power=0.9 npergroup=.;
run;
```

If this code is run, a sample size of 197 subjects per group is found.

6.7.2 Comparing Two Seroresponse Rates

Sample size estimation for a non-inferiority trial comparing two proportions and with the risk difference as effect measure is discussed in section 11.3.1 of the book by Julious [18]. Sample sizes from the preferred method, Method 1 (based on expected rates, the least conservative of the three methods), can be calculated with PROC POWER. With this procedure, sample sizes can be calculated for testing null hypotheses of the form

$$H_0 : \tilde{\pi}_B - \tilde{\pi}_A \leq d.$$

In case of a non-inferiority trial, the null hypothesis to test is

$$H_0 : \pi_{ref} - \pi_{test} \leq M.$$

Sample sizes for this null hypothesis can thus be estimated by setting $\tilde{\pi}_A$ to π_0, $\tilde{\pi}_B$ to π_1 and d to M.

An investigator wishes to estimate the sample size corresponding to a power of 0.9 for a non-inferiority trial with expected seroprotection rates $\pi_1 = 0.70$ and $\pi_0 = 0.75$ and non-inferiority margin $M = 0.1$, and with the null hypothesis being tested at the one-sided significance level 0.025. The required sample size can be obtained with SAS Code 6.4: 1,674 subjects per group.

A formula for sample size estimation for an equivalence trial comparing two proportions and with the risk difference as effect measure is formula (12.4) in the book by Julious [18]. SAS procedure PROC POWER does not contain a feature for equivalence trials with a rate as endpoint. But the formula is easy to program.

SAS Code 6.4 Sample size calculation for a non-inferiority trial with a seroprotection or a seroconversion rate as outcome

```
proc power;
   twosamplefreq test=pchi
   alpha=0.025 sides=1
   groupproportions=(0.75 0.70)
   nullpdifference=-0.10
   power=0.9 npergroup=.;
run;
```

6.7.3 Lot Consistency Trials

The standard approach to estimate the statistical power of a lot consistency design is to estimate separately the power of each of the three pair-wise comparisons, and then combine the resulting estimates to obtain an overall estimate.

Example 6.4 Ganju, Izu and Anemona investigate sample size estimation for vaccine lot consistency trials [38–40]. The example they use is a lot consistency trial for a quadrivalent vaccine for the prevention of meningococcal disease caused by *N. meningitidis* serogroups A, C, Y and W-135. For the C serogroup, they assume that $\sigma = \sqrt{6.15}$ for the $\log_2$-transformed antibody titres. They find that if there is no between-lot variation, i.e. if it is assumed that $\Delta_{12} = \Delta_{13} = 0$, that to be secured of an overall statistical power of 0.9, a sample size of $n = 500$ subjects per lot would be required (for $\alpha = 0.05$, and 1/1.5 to 1.5 as equivalence range). Because of the assumption that $\Delta_{12} = \Delta_{13} = 0$, each of the three single pair-wise comparison have the same power. This power can be calculated using SAS Code 6.2A with

```
meandiff=0 stddev=2.480
lowerr=-0.585 upper=0.585
alpha=0.05 npergroup=500
```

with $0.585 = \log_2 1.5$. The calculated power for the single pair-wise comparisons is 0.963. Using the Inequality (3.17), it follows that the overall statistical power for this lot consistency design is $\geq$0.889. If independence of the three pair-wise comparisons is assumed, then the overall statistical power is $0.963^3 = 0.893$.

Ganju and colleagues show that assumptions about Δ_{12} and Δ_{13} can have a profound impact on the statistical power of the design. They strongly argue against assuming that the between-lot variation is zero, and they advise to assume some amount of variation between lots. If non-zero between-lot variation is assumed, i.e. if it is assumed that $\Delta_{13} > 0$, then the statistical power will be highest for $\Delta_{12} = \Delta_{13}/2$

and lowest for $\Delta_{12} = 0$. The explanation for this is that equally spaced means are less variable than unequally spaced means and hence have a greater probability of demonstrating consistency.

Example 6.4 (continued) If Δ_{13} is set to 0.1 and Δ_{12} to 0.06, then with a lot size of 500 the power of the three pair-wise comparisons are: 0.949 (lot #1 versus lot #2), 0.923 (lot #1 versus lot #3) and 0.956 (lot #2 versus lot #3). These estimates can be obtained using in SAS Code 6.2B the statement

```
meandiff=0.1, 0.06, 0.04
```

Thus, now the overall statistical power would be

$$\geq (0.923 + 0.949 + 0.956) - 2 = 0.828,$$

or 0.837 if independence is assumed.

The standard approach to estimate the overall statistical power of a lot consistency design is slightly flawed, because it does not take the correlation between the three pair-wise comparisons into account. That the correlation between the comparisons is not zero is easy to see. If the mean differences between (*i*) lot #1 and lot #2 and (*ii*) lot #1 and lot #3 are known, then the mean difference between lot #2 and lot #3 is also known. Thus, the assumption that the three pair-wise comparisons are independent does not hold. Second, because the comparisons are dependent, separate estimation of the power of each of the three pair-wise comparisons also introduces bias. In practice, however, the amount of bias will be negligible, and the standard approach will give a good approximation of the actual power of the design.

An alternative method to estimate the overall power is Monte Carlo simulation. The algorithm for the simulation is as follows. A large number (≥5,000) of trials is simulated. Per trial, three random samples of size n are generated, with the data of the ith sample representing the log-transformed antibody values of the ith lot. The first random sample is drawn from a $N(0, \sigma^2)$ distribution, the second from a $N(\Delta_{12}, \sigma^2)$ distribution and the third from a $N(\Delta_{13}, \sigma^2)$ distribution. For each trial, the confidence intervals for the three pair-wise geometric mean ratios are calculated, and the trial is declared significant if all three confidence intervals fall in the predefined equivalence range. The statistical power of the design is then estimated as the proportion of simulated trials yielding a significant result. With this approach, the correlation between the comparisons is taken into account. For the example above, the simulated overall statistical power is 0.868.

Part III
Vaccine Field Studies

Chapter 7
Vaccine Field Studies

Abstract This chapter is an introductory chapter to vaccine field studies, i.e. vaccine efficacy and vaccine effectiveness studies. The difference between vaccine efficacy and vaccine effectiveness is explained, and the two major infection occurrence measures, attack rate and force of infection, are introduced. The formula that links the two measures, the so-called exponential formula, is explored in detail. The impact of the diagnostic test on vaccine efficacy and vaccine effectiveness estimates is inspected. The chapter is concluded with an investigation of the use of specific and non-specific endpoints in vaccine field studies. A formula is presented for the relationship between vaccine effectiveness against such endpoints and vaccine effectiveness against infection. It is shown that the general belief that a non-specific endpoint as surrogate for infection will always lead to underestimation of the vaccine effectiveness is false.

7.1 Protective Vaccine Effects

In vaccine field studies, protective effects of vaccines are investigated. Vaccines have two types of protective effects, direct effects and indirect effects. *Direct vaccine effects* are protective effects at subject level, effects in vaccinated subjects that are due to the vaccination and that are absent in unvaccinated subjects. *Indirect vaccine effects* are protective effects at population level that are due to a high percentage of the population having been vaccinated. In vaccine efficacy and vaccine effectiveness studies, direct vaccine effects are assessed.

Halloran, Struchiner and Longini present a theoretical framework for vaccine effects [41]. At subject level, historically the effect of interest has been the *vaccine effect for susceptibility to infection*. Here, the question is how well the vaccine protects vaccinated subjects against infection or disease, i.e. to what degree vaccination reduces the probability that a subject becomes infected or diseased if exposed to the pathogen. Standard endpoints in vaccine field studies for susceptibility to infection are occurrence of infection or disease and time to infection or disease.

Often, infection confers lifelong protection against the disease. This is, for instance, the case for mumps, measles and hepatitis A. For these diseases, a subject can get infected only once. This is, however, not true for all infectious diseases.

J. Nauta, *Statistics in Clinical and Observational Vaccine Studies*,
Springer Series in Pharmaceutical Statistics,
https://doi.org/10.1007/978-3-030-37693-2_7

Examples of diseases with possibly recurrent infections are acute otitis media (middle ear infection), genital herpes, meningitis (inflammation of the protective membranes covering the brain and spinal cord), cystitis (an inflammatory disorder of the bladder) and influenza. The reason why infection does not lead to lifelong protection is usually either that the naturally acquired antibodies against the pathogen do not offer sufficient protection, or exposure to serotypes of the pathogen that are not recognized by the antibodies. Cystic fibrosis (CF) is an inherited disease of the mucus. An abnormal gene causes the mucus to become thick and sticky. The mucus builds up in the lungs and blocks the airways, which makes it easy for bacteria to grow, leading to repeated, life-threatening lung infections. Over time, these infections can cause chronic progressive pulmonary disease, the most frequent cause of death in CF patients. The most prevalent of these infections is the one caused by the bacterium *Pseudomonas aeruginosa*. Naturally acquired antibodies against the bacterium often do not offer sufficient protection against the very virulent infections. A CF patient can become chronically colonized, with subsequent resistance to antibiotic courses. In a study with a *P. aeruginosa* vaccine in CF patients, measures of effect may thus be time to initial infection, time to colonization with *P. aeruginosa*, becoming chronically infected, number of recurrent infections or time between subsequent infections.

The *vaccine effect for progression or pathogenesis* is the protection the vaccination offers once a person has become infected. The vaccine may increase the incubation period, i.e. the time between infection and disease. Other effects of interest may be to what degree the vaccine reduces the intensity, the duration or the mortality from disease. A human immunodeficiency virus (HIV) vaccine may reduce the post-infection *viral load*, the amount of virus in body fluids. In HIV, keeping the viral load level as low as possible for as long as possible decreases the complications of the disease and prolongs life. Whereas the vaccine effect for susceptibility to infection requires that the study participants are free of infection at the time of their enrolment into the study, vaccine efficacy for progression or pathogenesis can only be studied in infected subjects.

A vaccinated subject may be less infectious to others, or he or she may be infectious for a shorter period of time. This is called the *vaccine effect for infectiousness*. These effects are of relevance at population level, because reduction of infectiousness has usually important health consequences. The effects will slow down the spread of the infection in the population. Vaccination of a large fraction of the population may lead to *herd immunity*. If a high percentage of a population immune is to an infection, then the spread of the infection may be prevented because it cannot find new hosts. Herd immunity gives protection to vulnerable members of the population such as newborns, the elderly and those who are too sick to be vaccinated. Herd immunity does not protect against all vaccine-preventable diseases, because not all infectious diseases are contagious. A *contagious* disease is one that spreads directly from person to person. An example of a non-contagious infectious disease is tetanus. It is caught from bacteria in the environment, not from people having the disease.

In this book, the focus is on vaccine efficacy and vaccine effectiveness for susceptibility to infection.

7.2 Vaccine Efficacy and Vaccine Effectiveness Studies

Vaccine field studies for susceptibility to infection are studies in which the occurrence of new cases of an infectious disease is compared between initially disease-free vaccinated and unvaccinated individuals. There are two types of field studies, vaccine efficacy studies and vaccine effectiveness studies. In everyday language, 'efficacy' and 'effectiveness' are often used as synonyms, but in the context of vaccine field studies the terms have distinctively different meanings. *Vaccine efficacy* is investigated in randomized, double-blind and controlled clinical studies, meaning that the protectiveness of the vaccine is measured under ideal conditions and in an often homogeneous population. Such a study is usually required for the approval by registration authorities of the vaccine. A major advantage of vaccine efficacy studies is rigour control of bias. A disadvantage of vaccine efficacy studies is lack of external validity, because it is not possible to predict accurately the level of protection that will be achieved in public health practice. A vaccine might be very protective in a homogenous population but be less protective in less homogenous and less selected populations. *Vaccine effectiveness* studies are observational studies, in which the protectiveness of a vaccine is investigated under real-world conditions, in a general population.

The effectiveness of a vaccine is less than its efficacy, and there are many factors contributing to this reduction in protectiveness. Vaccines need to be kept below certain temperatures to remain effective. To prevent them from becoming less protective when being transported, a so-called *cold chain system* is established. A cold chain system is a system that ensures that the vaccine is kept cool, from when it leaves the manufacturing site to when it is used. In clinical studies, the cold chain system is carefully monitored, and when vaccine supplies are accidentally exposed to temperatures outside the safe limits they are not used but destroyed. In the real world, the monitoring of the cold chain system is less strict, which will affect the vaccine's potency. Accidental freezing of vaccines has been long a largely overlooked problem, while freezing can also lead to potency loss.

Another example of a factor contributing to the reduced protectiveness is non-compliance with the vaccine dosing schedule. Some vaccines require multiple doses for optimal protection. Hepatitis vaccines require two or three doses. The human papillomavirus vaccine for adolescent and young adult women is recommended in a three-dose series. In efficacy studies, the compliance with multiple-dose vaccine schedules is usually satisfactory, but in the real world it is often suboptimal. Nelson et al. report that among those who received a first dose of a varicella, a hepatitis A or a hepatitis B vaccine, relatively low numbers completed the series: 55–65% for hepatitis B vaccine and 40–50% for hepatitis A and varicella vaccines [42]. They further note that among those who were compliant, often the interval between the doses was too long.

A third example is the exclusion of subjects with a weakened immune system due to having been treated with chemotherapy. In clinical vaccine trails, these subjects are usually excluded, but in the real world they are not (but they should not get any vaccines that contain live virus).

7.3 Definition of Vaccine Efficacy and Vaccine Effectiveness

Vaccine efficacy and *vaccine effectiveness* for susceptibility—the effect measures (parameters, quantities) being estimated in vaccine field studies—are defined as one minus the relative risk of infection for vaccinated subjects versus unvaccinated subjects:

$$\vartheta = 1 - \frac{\pi_1}{\pi_0}, \tag{7.1}$$

where ϑ is the vaccine efficacy/effectiveness, π_1 the risk of infection for vaccinated subjects and π_0 the risk for unvaccinated subjects. The smaller the relative risk of infection, the larger the vaccine efficacy or the vaccine effectiveness. ϑ will differ between the two types of studies (as said, usually smaller in vaccine effectiveness studies) but their definitions do not.

Example 7.1 If the risk of infection for unvaccinated subjects is 0.110 and that for vaccinated subjects 0.021, then the vaccine efficacy or, depending on the design of the field study, the vaccine effectiveness is

$$1 - \frac{0.021}{0.110} = 0.809.$$

or 80.9%.

Equation (7.1) can be rewritten as

$$\vartheta = \frac{\pi_0 - \pi_1}{\pi_0}.$$

This shows that vaccine efficacy and vaccine effectiveness are what epidemiologists call preventable fractions, the fractions of the risk of infection that can be directly prevented by vaccination. This is the formal definition of vaccine efficacy and vaccine effectiveness. A somewhat less formal definition is the proportion of infected cases directly preventable by vaccination. This explains why vaccine efficacy and vaccine effectiveness are often expressed as percentage.

The vaccine efficacy/effectiveness defined in (7.1) are properly called the *absolute* vaccine efficacy/effectiveness, absolute because the comparison is vaccinated versus unvaccinated. If the comparison is vaccinated with vaccine A versus vaccinated with vaccine B, then ϑ is called the *relative* vaccine efficacy or vaccine effectiveness, the relative change in the risk of infection. A relative vaccine efficacy or vaccine effectiveness can be negative.

Example 7.2 Assume that the risk of infection for subjects vaccinated with vaccine A is 0.340 and that for subjects vaccinated with Vaccine B 0.251. Then the relative vaccine efficacy/effectiveness of vaccine A to vaccine B is

$$1 - \frac{0.340}{0.251} = \frac{0.251 - 0.340}{0.251} = -0.355,$$

35.5% of the cases prevented by vaccine A will not be prevented by vaccine B.

7.4 Source Populations

7.4.1 Cohorts Versus Dynamic Populations

The set of individuals being monitored in a vaccine field study for the occurrence of cases of the infectious disease is called the source population (also study base). The two major types of source populations are cohorts and dynamic populations. It is, for many reasons, important to have a clear understanding of the difference between the two types of populations. To name just one, vaccine efficacy and vaccine effectiveness are defined only in reference to a cohort, while many vaccine effectiveness studies have a dynamic population as source population. This means that the results of these studies must somehow be 'translated' to a cohort. How this is done is explained in Sect. 7.8.2.

7.4.2 Cohorts

The definition of a cohort is a fixed group of initially infection-free but at-risk individuals, with the membership defined by an admissibility-defining event or condition, in a selected time period. What distinguishes a cohort from a dynamic population is that, once created, it is fixed with respect to its membership, and no new members can enter. Two examples of a cohort are

- all Finnish children aged 24–35 months at the start of an index influenza season and
- all subjects enrolled in a vaccine efficacy study.

The group of all Finnish children aged 24–35 months at the start of the influenza season constitutes a cohort. The condition for membership is living in Finland and being aged 24–35 months at the start of the influenza season. The admission period is the date of the start of the influenza season. Finnish children who will be two after the start of the season cannot become a member, because at the start of the season the membership became fixed. Members who turn three years of age after the start of the season, however, remain member. In the second example, the admissibility-defining event is enrolment into the study. Once the targeted number of participants has been reached, admission is stopped.

The start date of the follow-up (the surveillance period, see the next section) can be the same calendar date for all members of the cohort, but this is not a must. In the first example, the start date is the same for all members (start of the influenza season). But in vaccine efficacy studies, it is not uncommon that the follow-up starts immediately after the admission, in which case the start of the follow-up will be individual.

Ideally, at the time of admission to the cohort, for each member the vaccination status is fixed: vaccinated or not vaccinated. Otherwise, the vaccination status becomes time-dependent, which complicates the statistical analysis.

Example 7.3 Cholera is an acute diarrhoeal infection caused by ingestion of food or water contaminated with the bacterium *Vibrio cholerae*. The disease can cause acute diarrhoea with severe dehydration, which can lead to death if left untreated. Khatib et al. assessed the effectiveness of an oral cholera vaccine in Zanzibar [43]. The study population consisted of residents of a number of high-risk *shehias* (small administrative units) in the Zanzibar Archipelago. All healthy, non-pregnant residents aged two years and older were invited to participate in a mass vaccination campaign. Immunization (after verbal consent) with a killed whole-cell oral vaccine took place in January and February 2009. In this study, the source population is a cohort, with the membership defined as being a healthy, non-pregnant person aged two years and older, and being a resident of one of the selected shehias. The admission period was January–February 2009. At the start of the follow-up (March 2009) for all members, the vaccination status was fixed, and all members were at risk of contracting cholera. The monitoring period ended in May 2010.

It is often unavoidable that during the follow-up some members leave the cohort, by emigrating or death. If none of the members leaves the cohort, the cohort is said to be closed as to membership. If some of the members leave the cohort, by emigrating or death, it is said that the cohort is open.

7.4.3 Dynamic Populations

A dynamic population is a population at risk, in which the membership is not fixed but variable, with new members entering it and other members leaving it. An example of a dynamic population is the catchment population of a hospital. During the monitoring period, new members enter it by birth, by being referred by a GP, or by immigrating, while others leave it by death or emigrating. The vaccination status of the members need not to be fixed when they enter the population. A member can change vaccination status from not vaccinated when they join the population to vaccinated at some time later. This will, for example, be the case when sufficient vaccine supplies become only available after the outburst of a pandemic. A *pandemic* is a global outbreak that occurs when a new virus emerges for which people have little or no immunity. These disease spreads easily from person to person and can sweep around the world in very short time.

If at any period of time the number of vaccinated/unvaccinated members leaving the dynamic population is balanced by the number of vaccinated/unvaccinated new members entering it, it is said that the population is in steady state. Consider a study with a dynamic source population in which vaccination continues during the surveillance period. An unvaccinated member who gets vaccinated during the surveillance period is leaving the unvaccinated sub-population and enters the vaccinated sub-population. Because the fraction of vaccinated members increases during the surveillance period, such a population is not in steady state.

Example 7.4 Rotavirus is the leading cause of severe acute diarrhoea in children, both in developed and in developing countries, and it is the major cause of death in poor countries. Ichihara et al. evaluated the effectiveness of an oral attenuated monovalent vaccine, used in Brazil in routine health services, in preventing child hospitalization with acute diarrhoea [44]. They conducted a hospital-based case-control study in five regions of Brazil, from July 2008 to August 2011. Participating hospitals were general hospitals receiving children with a large range of diseases coming from a similar geographical area. Children were eligible to participate in the study if they were admitted to one of the study hospitals and were aged 4–24 months and therefore old enough to have received their second dose of rotavirus vaccine. In this study, the source population was dynamic, constituted by children aged 4–24 months and being a member of one of the catchment populations of the participating hospitals.

7.5 Surveillance Period

The *surveillance period* of a vaccine field study is the period or the time span during which the members of a source population are being monitored for the occurrence of new cases of the infectious disease.

If the study population is a dynamic population, the surveillance period is by definition a fixed calendar time interval.

Example 7.4 (continued) In the Brazilian rotavirus study, the source population was dynamic, and the surveillance period was from July 2008 to August 2011.

If the source population is a cohort, then the surveillance period is the length of time each member of the cohort is monitored for the occurrence of being infected. Thus, in case of a cohort, the duration a member is monitored is the same for all members. The surveillance period can but need not to be a fixed calendar time interval. This depends on the admission period. If the admission period is short, the surveillance will usually be a fixed calendar time interval, otherwise not.

Example 7.3 (continued) In this study, the admission period was relatively short, January–February 2009. The surveillance period started in March 2009 and ended in May 2010.

7.6 Risk of Infection

Risk of infection is defined as the probability that a non-infected individual becomes infected in a pre-specified time interval of fixed length, the *infection risk period*. During the whole risk period, the member should be at risk of becoming infected, unless infected.

7.7 Person-Time at Risk and Population-Time at Risk

An important concept in vaccine field studies is person-time at risk. An individual's person-time at risk is the time length he or she was a member of the source population while being disease-free. If, for example, the source population is a closed cohort, for members who did not become infected, the person-time at risk is the length of the surveillance period. For a member who did become infected, the person-time at risk is the length of the period between the start of the surveillance and the infection. The unit for measuring person-time at risk is arbitrary, it can be person-years or person-months, or person-weeks, arbitrary because 1.5 person-years is equal to 18 person-months and vice versa. The sum of the individual person-times at risk of the members of the source population is called the *population-time at risk* or the *total person-time at risk.*

7.8 Infection Occurrence Measures

7.8.1 Attack Rate and Force of Infection

The parameters to be estimated in a vaccine field study are either attack rates or forces of infection. The *attack rate* is the expected fraction of a cohort at risk that becomes infected during the surveillance period. The link to the risk of infection for members of the cohort is that the attack rate can be interpreted as the average risk of infection. Although attack rate applies to a cohort and risk of infection to individuals, the two terms are often used synonymously. For ease of readability, risk of infection will be used instead of the more accurate average risk of infection.

The *force of infection* of an infectious disease is the expected number of new cases of the disease per unit person-time at risk. Force of infection applies to a source population, not to individuals. Of the two measures of infection occurrence, force of infection is the most fundamental one. It is usually assumed that the occurrence of infected cases is a homogeneous Poisson process, see Appendix E. A homogenous Poisson process is determined by a single parameter, its intensity λ. One property of the process is that $\mathbf{N}(L)$, the number of events in any interval of length L, is a Poisson random variable with parameter (or mean, expectation) λL:

$$\mathrm{E}\{\mathbf{N}(L)\} = \lambda L.$$

When the events are infected cases, λ is called the force of infection, and L is then the total person-time at risk. Thus, if the expected number of infected cases is 18 cases per 1,000 total person-weeks at risk, then the force of infection is 18/1,000 cases per person-week. As said, the time unit is arbitrary, because 1 case per person-week at risk is the same as 52 cases per person-year at risk, and vice versa.

Example 7.5 Assume that in a dynamic population the force of infection is 1.4 infected subjects per 100 months person-time, and that the population is being monitored for one year, leading to a total population-time-at risk of, say, 80,000 months. Then the expected number of new cases to occur in the population during the surveillance period is

$$\frac{80{,}000 \text{ months} \times 1.4 \text{ cases}}{100 \text{ months}} = 1{,}120 \text{ cases.}$$

A force of infection need not be to be homogeneous or uniform; it can also be non-homogeneous, a function of time. If the force of infection is non-homogeneous, $\lambda(t)$ is called the *force of infection function*. Here, t is time since the start of the surveillance period, not person-time at risk.

7.8.2 Link Between the Risk of Infection and the Force of Infection

Because vaccine efficacy and vaccine effectiveness are defined in terms of risks of infection (see Sect. 7.3), it would be convenient if there was a formula linking risk of infection and force of infection. Fortunately, such a formula exists. They are linked by the following formula [45, 46]:

$$\pi(t_s) = 1 - \exp\left[-\int_0^{t_s} \lambda(u)\, du\right]. \tag{7.2}$$

In the epidemiology this formula, cited in many textbooks, is sometimes referred to as 'the exponential formula linking risk and incidence', or simply as 'the exponential formula' [47]. In Appendix E, a derivation of the formula is presented.

Example 7.5 (continued) Assume that in the cohort study the duration of the surveillance period is 18 months. Then the expected attack rate is

$$\begin{aligned}\pi &= 1 - \exp\left[-\int_0^{18} 1.4/100\, du\right] \\ &= 1 - \exp[-18 \times 1.4/100] \\ &= 1 - 0.777 = 0.223.\end{aligned}$$

The integral on the right-hand side of Eq. (7.2) is called the *cumulative force of infection function*:

$$\Lambda(t_s) = \int_0^{t_s} \lambda(u)\, du.$$

It is possible to derive a simpler and more convenient version of Form. (7.2). Remember that if $x > 0$ is small

$$\exp(-x) \approx 1 - x,$$

and thus

$$1 - \exp(-x) \approx x.$$

Hence, when $\Lambda(t_s)$ is small

$$1 - \exp[-\Lambda(t_s)] \approx \Lambda(t_s),$$

and Form. (7.2) becomes

$$\pi(t_s) \approx \Lambda(t_s). \tag{7.3}$$

The approximation works well when $\pi(t_s)$ is smaller than 0.10.

If the force of infection is homogeneous, Form. (7.3) further simplifies to

$$\pi(t_s) \approx \lambda \times t_s. \tag{7.4}$$

In the epidemiology this is often expressed as

$$\text{Risk} \approx \text{Incidence} \times \text{Time},$$

see, for example, Rothman [48].

Form. (7.2) holds only for cohorts, but not for dynamic populations (although a similar formula holds for dynamic populations that are in steady state, see Appendix E). This is because risk of infection is only defined in reference to a cohort. But what if a force of infection is estimated from data of a study in a dynamic population, which is, see Chap. 9, often the case? In that case the formula is applied to a hypothetical cohort, constituted by individuals whose risk profiles are similar to those of the members of the dynamic population. This is less far-fetched than it perhaps seems, because it is often not so difficult to turn a dynamic population into a cohort, by slightly modifying the admission criteria. The surveillance period, however, should not be too long, because during the surveillance period a cohort ages, while dynamic populations do not, or at a slower speed. In other words, the effect of ageing on the force of infection is not measurable in a dynamic population.

7.8.3 Recapitulation

The previous section can be summarized as follows:

- The two measures for the occurrence of infection are the attack rate (average risk of infection) and the force of infection.
- The link between the risk of infection and the force of infection is the exponential formula.

- When the risk of infection is smaller than 0.10, the risk of infection is approximately equal to the cumulative force of infection, i.e. the total force of infection during the surveillance period.
- When the force of infection is homogeneous, the risk of infection is approximately equal to the force of infection times the duration of the surveillance period.

7.8.4 When the Relative Risk of Infection Cannot be Estimated

This might come as a surprise, but only a few vaccine field study designs allow the estimation of the relative risk of infection, and thus of the vaccine efficacy or the vaccine effectiveness, and even fewer designs—in fact only one—allow the relative risk to be estimated by means of efficient methods like regression analysis. In most vaccine field study designs, what can be estimated is the relative force of infection.

The relative risk of infection can be estimated only in designs with a cohort as source population, in so-called cohort studies. If the cohort is closed, estimation is straightforward, and the data can be analysed by means of relative risk regression (see Chap. 9). When the cohort is open, it still is possible to estimate the risks of infection, with the Nelson–Aalen estimator (see Sect. 8.2.4), and thus the relative risk, but a corresponding regression approach is not available. What can be estimated efficiently when the cohort is open is the relative force of infection, by means of Poisson regression or Cox regression (see Sects. 8.2.2 and 8.2.3).

When the source population is dynamic, the parameter that can be estimated is the *relative force of infection*, which is defined as

$$\phi = \frac{\Lambda_1(t_s)}{\Lambda_0(t_s)},$$

the ratio of the two cumulative forces of infection. From (7.3), it follows that

$$\phi \approx \frac{\pi_1(t_s)}{\pi_0(t_s)},$$

and thus that

$$\phi \approx \theta.$$

Thus the relative force of infection ϕ is approximately equal to the relative risk of infection θ. Denote

$$\vartheta_\phi = 1 - \frac{\Lambda_1(t_s)}{\Lambda_0(t_s)},$$

then

$$\vartheta_\phi \approx \vartheta, \tag{7.5}$$

Table 7.1 ϑ_ϕ by π_0 and ϑ

ϑ					
π_0	0.10	0.30	0.50	0.70	0.90
0.05	0.10	0.31	0.51	0.71	0.90
0.10	0.10	0.31	0.51	0.71	0.90
0.15	0.11	0.32	0.52	0.72	0.91
0.20	0.11	0.32	0.53	0.72	0.91

and an estimate of ϑ_ϕ will be an approximate estimate of ϑ. Thus, when the study design allows estimation of the relative force infection but not of the relative risk, it is still possible to obtain an approximate estimate of the vaccine efficacy/effectiveness.

There remains the question, for what values for π_0 and π_1 does (7.5) hold? A first guess would be when both π_0 and π_1 are smaller than 0.10 (see Sect. 7.8.2). Surprisingly, this condition is much too strict. In Table 7.1, values for ϑ_ϕ are shown for $\pi_0 = 0.05, 0.10, 0.15$ and 0.20, and $\vartheta = 0.10, 0.30, 0.50, 0.70$ and 0.90. For example, when $\pi_0 = 0.15$ and $\vartheta = 0.70$, then

$$\pi_1 = (1 - \vartheta)\pi_0 = 0.045,$$

$$\Lambda_0 = -\log(1 - \pi_0) = 0.163,$$

$$\Lambda_1 = -\log(1 - \pi_1) = 0.046,$$

and

$$\vartheta_\phi = 1 - \frac{\Lambda_1}{\Lambda_0} = 0.718.$$

Thus, if $\pi_1 < \pi_0$, then $\vartheta_\phi \geq \vartheta$, but when the risk of infection among unvaccinated subjects is less than 0.20, the difference between ϑ_ϕ and ϑ is negligible for all practical purposes.

7.9 Four Simulated Cohort Studies

To make the concept force of infection and the formula linking it to the risk of infection less abstract, data of four cohort studies were simulated. These data are analysed in Chap. 8. The SAS code to reproduce the data set COHORTS is given in Appendix H.

In each cohort, the number of vaccinated members is set to $n_1 = 4{,}000$, the number of unvaccinated members to $n_0 = 5{,}000$ and the length of the surveillance period t_s to 140 weeks. Both homogeneous and non-homogeneous forces of infections are assumed. The homogeneous forces of infection are $\lambda_1 = 0.8$ case per 1,000 person-weeks in the vaccinated group, and $\lambda_0 = 1.2$ case per 1,000 person-weeks in the unvaccinated group. The non-homogeneous forces of infection (cases per 1,000 person-weeks) are

$$\lambda_1(t) = \begin{cases} 1.2 & t < 40 \text{ weeks} \\ 1.0 & \text{if} \quad 40 \text{ weeks} \leq t < 80 \text{ weeks} \\ 0.7 & t \geq 80 \text{ weeks} \end{cases}$$

for the vaccinated group, and

$$\lambda_0(t) = \begin{cases} 2.4 & t < 40 \text{ weeks} \\ 2.0 & \text{if} \quad 40 \text{ weeks} \leq t < 80 \text{ weeks} \\ 1.4 & t \geq 80 \text{ weeks} \end{cases}$$

for the unvaccinated group. Note that with these choices the two non-homogeneous forces of infection are proportional. For both combinations of forces of infections, two cohort versions are simulated, a closed and an open version.

For the unvaccinated group, the cumulative homogeneous force of infection is

$$\begin{aligned} \Lambda_0 &= \int_0^{140} 1.2/1{,}000 \, du \\ &= 140 \times 0.0012 = 0.168, \end{aligned}$$

which corresponds to a risk of infection (i.e. an expected attack rate) of

$$\pi_0 = 1 - \exp(-0.168) = 0.155.$$

The cumulative force of infection in the vaccinated group is

$$\begin{aligned} \Lambda_1 &= \int_0^{140} 0.8/1{,}000 \, du \\ &= 140 \times 0.0008 = 0.112, \end{aligned}$$

and the risk of infection

$$\pi_1 = 1 - \exp(-0.112) = 0.106.$$

Thus, the relative risk of infection is

$$\theta = \frac{0.106}{0.155} = 0.684, \tag{7.6}$$

and the relative cumulative force of infection is

$$\begin{aligned}\phi &= \frac{\Lambda_1}{\Lambda_0} \\ &= \frac{0.112}{0.168} = 0.667.\end{aligned} \tag{7.7}$$

Hence

$$\vartheta = 0.32 \quad \text{and} \quad \vartheta_\phi = 0.33.$$

For the unvaccinated group, the cumulative non-homogeneous force of infection is

$$\Lambda_0 = 40 \times 0.0024 + 40 \times 0.0020 + 60 \times 0.0014 = 0.260,$$

and $\pi_0 = 0.229$. The cumulative non-homogeneous force of infection in the vaccinated group is

$$\Lambda_1 = 40 \times 0.0012 + 40 \times 0.0010 + 60 \times 0.0007 = 0.130,$$

and the risk of infection $\pi_1 = 0.122$. Thus, the relative risk of infection is

$$\theta = \frac{0.122}{0.229} = 0.533, \tag{7.8}$$

and the relative cumulative force of infection is

$$\phi = \frac{0.130}{0.260} = 0.500. \tag{7.9}$$

Thus

$$\vartheta = 0.47 \quad \text{and} \quad \vartheta_\phi = 0.50.$$

The relative large difference between ϑ and ϑ_ϕ is explained by π_0 being greater than 0.20 (cf. Sect. 7.8.4).

7.10 Impact of the Diagnostic Test on the Vaccine Efficacy or Effectiveness Estimate

One of the most critical aspects of vaccine field studies is case definition. First, it has to be decided whether infection or disease is the endpoint of interest. This choice is of importance because infection does not necessarily imply developing the disease. As a rule of the thumb, vaccine field studies with disease as endpoint require larger sample sizes and a longer surveillance period than studies with infection as endpoint. On the other hand, case finding may be easier with disease as endpoint. Disease is usually accompanied by specific clinical symptoms, while infection requires laboratory confirmation. With infections with a long incubation period, such as HIV, for example, infection as endpoint would require repeated laboratory testing, which sometimes is difficult to organize (study participants having to visit the investigational site at predefined times, etc.) and can be very costly (while most laboratory results will be negative). If the incubation period is short, like in the case of pertussis, laboratory testing is often done after observing clinical symptoms of the disease. A drawback of this case-finding strategy is that *asymptomatic infections*, which are infections without clinical symptoms, go undetected. In case of influenza, culture confirmation is only possible during the first two to three days after infection. If the culture specimen collection is done too late, the infection will go undetected. It is sometimes argued that it is disease that matters, not infection. That is a too hasty conclusion. Infections do not only cause the disease, some can also do damage to organs. Asymptomatic infections should not be considered as being without risk. Sexually transmitted infections, in particular, are known for not producing clinical symptoms. If treatment for the infection is delayed or never given, this can cause permanent damage to the reproductive organs. In fact, almost any type of infection can impair fertility, in particular, those that affect the reproductive tract, including the prostate, the epididymis or the testis. A harmless infection such as the common cold may temporarily lower the sperm count.

Case finding requires a diagnostic test. (Diagnostic test is used here in the broadest sense of the word. It can be a single clinical or laboratory test but it can also be a diagnostic strategy, for example, laboratory testing only after the manifestation of certain clinical symptoms.) If the test misclassifies non-cases as cases of the infection or the disease, the vaccine efficacy or effectiveness estimate will be biased towards null. If, on the other hand, the test misclassifies cases as non-cases, this will not necessarily bias the efficacy or effectiveness estimate, but it may. If the diagnostic test detects moderate to severe cases easier than mild cases, and if vaccinated cases are milder than placebo cases, the vaccine efficacy or effectiveness estimate will be biased towards one.

Diagnostic tests are rarely totally accurate, and a proportion of the cases will be misclassified as non-cases. Such cases are called false-negative cases. The proportion of misclassified cases will depend on a property of the diagnostic test known as the sensitivity. The sensitivity of a diagnostic test is the conditional probability that the test will be positive (Test+) if the disease is present (Disease+):

$$\text{sensitivity} = \Pr(\text{Test+} \mid \text{Disease+}).$$

The false-negative rate is equal to one minus the sensitivity (1-sensitivity).

Not only may cases be misclassified as non-cases, non-cases may be classified as cases. Such cases are called false-positive cases and their number will depend on the specificity of the test. The specificity of a diagnostic test is the conditional probability that the test will be negative (Test−) given that the disease is absent (Disease−):

$$\text{specificity} = \Pr(\text{Test}- \mid \text{Disease}-).$$

The false-positive rate is equal to one minus the specificity (1-specificity).

Example 7.6 (*i*) Dengue fever, also known as breakbone fever, is a viral infection transmitted by mosquitoes. The disease is endemic to almost all tropical and subtropical regions. The disease causes flu-like symptoms, which in some cases can become life-threatening. Assume that in a randomized vaccine field study, 1,000 elderly are vaccinated with a dengue vaccine and a further 500 with placebo, and that during the surveillance period 70 members of the cohort contract the disease, 25 in the vaccine group and 45 in the placebo group. Thus, the vaccine efficacy for the prevention of dengue fever is

$$VE = \frac{45/500 - 25/1{,}000}{45/500} = 0.722.$$

Assume that the sensitivity of the diagnostic test for the fever is not perfect, not 1.0, but, say, 0.8. In that case, the (expected) number of false-negative cases will be

$$(1.0 - 0.8) \times 70 = 14,$$

with 9 misclassified cases in the placebo group and 5 in the vaccinated group. But the vaccine efficacy would still be correctly estimated:

$$\begin{aligned} VE &= \frac{(45-9)/500 - (25-5)/1{,}000}{(45-9)/500} \\ &= \frac{(0.8)45/500 - (0.8)25/1{,}000}{(0.8)45/500} \\ &= \frac{45/500 - 25/1{,}000}{45/500} = 0.722. \end{aligned}$$

(*ii*) Next, assume that the specificity of the test is less than 1.0, say, 0.98. Then, of the 1,430 subjects who did not contract the fever,

$$0.02 \times 1,430 = 28.6$$

will test positive, 9.1 in the placebo group and 19.5 in the vaccinated group. The vaccine efficacy would be estimated to be

$$VE = \frac{(45 + 9.1)/500 - (25 + 19.5)/1{,}000}{(45 + 9.1)/500} = 0.589$$

which is indeed a (considerable) bias towards null. Finally, it may be that the diagnostic test is such that less severe cases go undetected, and that the less severe cases are predominantly in the vaccine group. In that case, the vaccine efficacy will be overestimated.

(*iii*) Assume that the sensitivity of the diagnostic test for dengue fever is 1.0 for severe cases but only 0.9 for non-severe cases, and that in the placebo group 30% of the cases are non-severe but in the vaccine group 60%. In that case, in the placebo group

$$(1.0 - 0.9) \times 0.3 \times 45 = 1.35$$

cases will be misclassified as non-cases, while in the vaccine group

$$(1.0 - 0.9) \times 0.6 \times 25 = 1.5$$

cases will be misclassified. The vaccine efficacy would be estimated to be

$$VE = \frac{(45 - 1.35)/500 - (25 - 1.5)/1{,}000}{(45 - 1.35)/500} = 0.731,$$

a modest overestimation of the vaccine efficacy.

7.11 Specific and Non-specific Endpoints

In vaccine effectiveness (and efficacy) studies, both specific and non-specific endpoints are employed. A *specific endpoint* is based on a laboratory test for virus detection, for example, the polymerase chain reaction (PCR) test, which can detect even the smallest amount of virus DNA. A *non-specific endpoint* refers to an infection-related but not laboratory-confirmed clinical syndrome. (An endpoint is *infection-related* if infected subjects are more susceptible to it than non-infected subjects.) An example of such a syndrome is influenza-like illness (ILI), a respiratory illness characterized by fever, fatigue, cough and other symptoms. Influenza is not the only cause of ILI, it can be caused by other viruses as well, for example, by the norovirus (the 'winter vomiting bug', an outbreak of the virus occurred in Pyeongchang ahead of the Winter Olympics 2018) the most common cause of viral gastroenteritis. Because influenza is not the only cause of ILI, the syndrome is non-specific for influenza.

Non-specific endpoints in vaccine effectiveness studies are used for two reasons. First, ease of case finding, as usually non-specific endpoint cases are easier to identify than specific endpoint cases. ILI is easy to diagnose clinically, while for the diagnosis of influenza laboratory testing is required. Second, the endpoint may be serious or life-threatening, and an investigator may want to assess the impact of influenza vaccination on its occurrence. An example of a serious endpoint is myocarditis (inflammation of the heart muscle), most often caused by a viral infection.

A laboratory test for infection will usually only be done in case of clinical signs and symptoms, which means that asymptomatic infections will go undetected. Furthermore, when the PCR test is used to determine infection, a nasopharyngeal swab must be collected. It may not be possible to collect all specimens within a few days of symptoms onset and the laboratory test may be done too late, after viral shedding has stopped, causing a number of cases of the infectious disease to be missed. This means that the sensitivity of the PCR test is not perfect. But this will not cause bias, see Example 7.6.

For the relationship between vaccine effectiveness against specific and non-specific endpoints and vaccine effectiveness against infection, a formula exists, the *Nauta–Beyer formula* [49]. In its simplest form, the formula is

$$\vartheta_E = \frac{(\theta_E - 1)\vartheta_I}{(\theta_E - 1) + 1/\pi_u}, \tag{7.10}$$

where ϑ_E is the vaccine effectiveness against the endpoint, ϑ_I the vaccine effectiveness against the infection, θ_E the relative risk of the endpoint for *infected* subjects versus *non-infected* subjects, and π_u the infection attack rate in unvaccinated subjects. The relative risk θ_E is a measure of the strength of the relationship between the endpoint and infection. The relationship is strong for specific endpoints and weak for non-specific endpoints. If there is no relationship between the endpoint and infection, like for the endpoint toothache, for example, θ_E equals 1.0.

Example 7.7 Consider a non-specific endpoint for which influenza-infected subjects are 8 times more at risk than non-infected subjects. Then $\theta_E = 8.0$. Assume that ϑ_I, the vaccine effectiveness against influenza, is 0.65. Then, in seasons with a low influenza attack rate among unvaccinated subjects, say, 0.02, the vaccine effectiveness against the endpoint will be $\vartheta_E = 0.080$. In seasons with a high attack rate, say, 0.10, the vaccine effectiveness against the effectiveness will be $\vartheta_E = 0.268$.

From (7.10) it follows that

- the vaccine effectiveness against an endpoint depends on the infection attack rate in unvaccinated subjects;
- the higher the attack rate, the higher the vaccine effectiveness against the endpoint;
- the stronger the strength between the endpoint and infection, the less impact the attack rate has.

Example 7.8 Consider the use of ILI as surrogate endpoint for symptomatic influenza infection. Because by definition all symptomatic influenza cases develop ILI symptoms, the risk of ILI for influenza-infected subjects is 1.0. Jackson et al. find that the risk of ILI for subjects who do not get infected by influenza is approximately 0.097 [50]. Thus,

$$\theta_{ILI} = \frac{1}{0.097} \approx 10.0,$$

a value implying that the strength of the relationship between ILI and symptomatic influenza infection is only modest. If $\vartheta_I = 0.70$ and the influenza attack rate between seasons varies between 0.01 and 0.10, then, according to Form. (7.10), ϑ_E will vary between 0.058 and 0.332, due to this variation in the attack rate.

Assume that the endpoint is non-specific. In a vaccine effectiveness study, it is usually not the occurrence of the endpoint (the clinical syndrome) that is observed but rather the result of a diagnostic test for the clinical syndrome. If this is taken into account, the 'risk of the endpoint' becomes the risk of a positive test result. For infected subjects, this risk is

$$\Pr(\text{Test+} \mid \text{I+}),$$

and the risk for non-infected subjects is

$$\Pr(\text{Test+} \mid \text{I}-) = 1 - \Pr(\text{Test}- \mid \text{I}-).$$

These risks can be interpreted as the sensitivity (*sens*) and (one minus) the specificity (*spec*) of the diagnostic test, not for the detection of the clinical symptom but for the detection of the infection. Then

$$\theta_E = \frac{sens}{1 - spec}.$$

If the specificity of the test is perfect, then $\vartheta_E = \vartheta_I$, independent of the sensitivity of the test, because then

$$\theta_E - 1 = \infty,$$

implying that

$$\frac{1}{(\theta_E - 1)\pi_u} = 0.$$

This is the proof of the claim made in Example 7.6 (*i*).

Example 7.6 (*ii*) (continued) Here $\vartheta_I = 0.722$, the infection attack rate in the unvaccinated subjects is $45/500 = 0.09$, $sens = 1.0$ and $spec = 0.98$ and thus $\theta_E = 50$, and

$$\vartheta_E = \frac{49 \times 0.722}{49 + 1/0.09} = 0.589,$$

the same value as found earlier.

Form. (7.10) rests on two assumptions that the endpoint risk does not differ between neither

1. vaccinated and unvaccinated *non-infected* subjects nor
2. vaccinated and unvaccinated *infected* subjects.

Because vaccination state is not likely to influence the risk of developing the endpoint due to other pathogens than the virus of interest, the first assumption will usually be met. The second assumption will not be met if vaccinated infected subjects become less clinically ill from the infection, making them less susceptible to the endpoint compared to unvaccinated infected subjects. To express this difference in susceptibility, the symbol γ is used. If, for example, $\gamma = 0.8$, then the risk of the endpoint for *vaccinated* infected subjects is 0.8 times the risk for *unvaccinated* infected subjects. The full formula for the relationship between ϑ_E and ϑ_I then is

$$\vartheta_E = \frac{(\theta_{Eu} - 1) - (\gamma\theta_{Eu} - 1)(1 - \vartheta_I)}{(\theta_{Eu} - 1) + 1/\pi_u}, \tag{7.11}$$

where θ_{Eu} is the relative risk of the endpoint for infected *unvaccinated* subjects versus non-infected *unvaccinated* subjects.

Example 7.7 Assume that $\vartheta_I = 0.7, \theta_{Eu} = 9.0, \pi_u = 0.05$ and $\gamma = 1.0$. Then $\vartheta_E = 0.2$. Next, assume that $\gamma = 0.6$. Then $\vartheta_E = 0.24$. This difference is explained by the fact that vaccinated endpoint cases are not only prevented by the vaccination but also by being less susceptible to the endpoint when infected.

Assume again that the endpoint is a diagnostic test rather than a clinical syndrome. Let the specificity of the test be perfect. With a little algebra, it may be shown that then

$$\vartheta_E = 1 - \gamma + \gamma\vartheta_I, \tag{7.12}$$

where γ is now the ratio of the sensitivity of the test in vaccinated subjects to the sensitivity of the test in unvaccinated subjects.

Example 7.6 (*iii*) (continued) The sensitive of the test in unvaccinated subjects is

$$1.0(0.7) + 0.9(0.3) = 0.97,$$

and the sensitivity of the test in vaccinated subjects is

$$1.0(0.4) + 0.9(0.6) = 0.94.$$

Hence

$$\gamma = \frac{0.94}{0.97} = 0.969,$$

and

$$\vartheta_E = 1 - 0.969 + 0.969(0.722) = 0.731.$$

From Eq. (7.12), it follows that the vaccine effectiveness will be overestimated when $\gamma < 1$, i.e. when the sensitivity of the test in vaccinated subjects is less than the sensitivity of the test in unvaccinated subjects.

If in a vaccine effectiveness study, a non-specific endpoint is employed not as a surrogate for infection, but because the endpoint is serious or life-threatening, the question arises how to interpret the resulting effectiveness estimate. A major but often overlooked difference between vaccine effectiveness against infection and vaccine effectiveness against a non-specific endpoint is that the maximum value for ϑ_I is known (1.0), while the maximum value for ϑ_E is unknown and less than 1.0, due to its dependency on the attack rate π_u. This has considerable consequences for the interpretation of ϑ_E estimates. ϑ_I estimates are easy to interpret: a value of 0.35 means a moderate effectiveness against infection (only 35% of infected cases prevented), given the maximum possible value of 1.0 (100% of infected cases prevented). For ϑ_E, such univocal interpretation in terms of low, moderate or high is less straightforward. An estimate of 0.15 means 15% of endpoint cases prevented, but it is often difficult to decide if this is a low or high effectiveness. To be able to use the adjectives low or high, the 'costs' of an endpoint case have to be taken into account. For some endpoints, for example, death from all causes, 5% of cases prevented may be considered an excellent vaccine effectiveness, while for other endpoints it may imply a poor effectiveness.

Another way of looking at it is to relate ϑ_E to its maximum possible value. Assume that Form. (7.10) applies. Then the maximum possible value for ϑ_E is

$$\vartheta_E = \frac{(\theta_E - 1)}{(\theta_E - 1) + 1/\pi_u},$$

when $\vartheta_I = 1$. It then follows that

$$\vartheta'_E = \frac{\vartheta_E}{\text{maximum value}} = \vartheta_I.$$

In other words, when the vaccine effectiveness against a non-specific endpoint is related to its maximum possible value, then the vaccine effectiveness against the endpoint is equal to the vaccine effectiveness against infection.

Chapter 8
Vaccine Efficacy Studies

Abstract This chapter focuses on the analysis of vaccine efficacy data. Special attention is given to the analysis of person-time or time-to-infection data. Four different approaches are presented, an exact one, two asymptotic ones (Poisson regression and Cox regression) and a new one, based on bootstrapping. The assumptions underlying them are inspected. Further, a new vaccine efficacy estimator is introduced here for the first time. This estimator does not require any distributional assumptions and is based on an estimator that is quickly gaining popularity, the Nelson–Allen estimator.

8.1 Comparing Attack Rates

In a vaccine efficacy study, the source population is a cohort, constituted by the subjects enrolled in the study. If the cohort is closed, the vaccine efficacy analysis involves comparing the observed attack rate R_1 in the vaccinated group with the observed attack rate R_0 in the unvaccinated group. The estimator VE of the vaccine efficacy ϑ is

$$
\begin{aligned}
VE &= 1 - RR \\
&= 1 - \frac{R_1}{R_0} = 1 - \frac{c_1/n_1}{c_0/n_0},
\end{aligned}
$$

where RR is the attack rate ratio, c_1 and c_0 the numbers of cases (infected subjects) that occurred during the surveillance period, and n_1 and n_0 the group sizes. If

$$(LCL_\theta, UCL_\theta)$$

is a two-sided $100(1-\alpha)\%$ confidence interval for the relative risk of infection θ, then

$$(1\text{-}UCL_\theta, 1\text{-}LCL_\theta)$$

is a two-sided $100(1-\alpha)\%$ confidence interval for ϑ, the vaccine efficacy.

J. Nauta, *Statistics in Clinical and Observational Vaccine Studies*,
Springer Series in Pharmaceutical Statistics,
https://doi.org/10.1007/978-3-030-37693-2_8

Example 8.1 Recall the simulated cohort data discussed in Sect. 7.9. In the SAS data set `Cohorts` (see Appendix H), for the closed cohort with the non-homogeneous forces of infection, $c_1 = 486$ and $c_0 = 1{,}150$. Thus, $R_1 = 0.1215$, $R_0 = 0.230$ and

SAS Code 8.1 Comparing two attack rates

```
proc sort data=Cohorts out=Dsort;
   by descending Group descending Case;

proc freq data=Dsort order=data;
   table Group*Case / relrisk;
   where Version="closed" & FOI="non-homogeneous";
run;
```

$$RR = \frac{1,150/4,000}{486/5,000} = 0.528.$$

The data can be analysed using SAS Code 8.1. As two-sided 95% confidence interval for the relative risk θ

$$(0.479, 0.582)$$

is returned, which contains the true value of θ, 0.533 (see Eq. (7.8)).

Occasionally, it will not be sufficient to demonstrate that the vaccine efficacy is greater than zero, but that the requirement is that it has to be shown that the efficacy is substantially greater than zero. This is called *super efficacy*. For example, for influenza vaccines, the FDA/CBER requirement is that the vaccine efficacy must be greater than 0.4 [51]. This is to be demonstrated by showing that the lower limit of the two-sided 95% confidence interval for the vaccine efficacy exceeds 0.4.

Example 8.2 Blennow et al. report the result of a whole-cell pertussis vaccine efficacy study [52]. The study was performed in Sweden, in the early nineteen-eighties. In this non-blinded study, 525 infants aged 2 months who were born on days with an even number received three doses of vaccine one month apart, and 615 infants of the same age who were born on days with an odd number were enrolled as controls. The surveillance period was the age period between 6 and 23 months. In the vaccinated group, 8 cases of pertussis occurred, compared to 47 in the control group. The estimated absolute vaccine efficacy was

$$VE = 1 - \frac{8/525}{47/615} = 0.801.$$

In the vaccinated group, 80% of the expected cases of pertussis were prevented. The two-sided 95% Wilson-type confidence interval for the relative risk θ is

$$(0.097, 0.410),$$

which corresponds to a 95% confidence interval for the vaccine efficacy ϑ of

$$(0.590, 0.903).$$

Assume that there had been the requirement of super efficacy, with the requirement being that the vaccine efficacy exceeds 0.5. Because the lower limit of the confidence interval for vaccine efficacy falls above this limit, the requirement would have been met.

8.2 Comparing Person-Time Data

When the cohort is open, the data to compare between the vaccinated and unvaccinated members are person-time data. For the infected members of the cohort, the person-time is the time until the infection. For the non-infected members, the person-time is the length of the surveillance period; for non-infected members who leave the cohort, the person-time is the duration of the membership.

There are several statistical methods to compare person-time data between groups. Below, four are discussed, an exact one, two asymptotic ones and one based on bootstrapping. The first method requires the assumption that the two underlying forces of infection are homogeneous. The second method does not have this requirement. The third method also does not require homogeneity, but the force of infection functions must be proportional. The fourth method, the bootstrap method, does not require any distributional assumption.

8.2.1 An Exact Conditional Test

When it can be assumed that the force of infection λ is homogeneous, the estimator to use is the *infection incidence rate*:

$$IR = \frac{c}{TPT},$$

where c is the total number of infected members of the cohort (cases) and TPT the total person-time. IR is an unbiased estimator of λ. The effect measure is the *infection incidence rate ratio*, the ratio of the infection incidence rate IR_1 in the vaccinated group and the incidence rate IR_0 in the unvaccinated group:

$$IRR = \frac{IR_1}{IR_0}.$$

IRR is an estimator of the ratio of the two forces of infection λ_1 and λ_0, and of the ratio ϕ of the two cumulative forces of infection Λ_1 and Λ_0, because

$$\phi = \frac{\Lambda_1}{\Lambda_0} = \frac{\lambda_1 t_s}{\lambda_0 t_s} = \frac{\lambda_1}{\lambda_0}.$$

The vaccine efficacy that can be estimated is ϑ_ϕ (see Sect. 7.8.4), and the estimator is

$$VE_\phi = 1 - IRR.$$

Example 8.1 (continued) The numbers of cases in the open cohort with homogeneous forces of infection are $c_1 = 411$ and $c_0 = 759$. The total person-times are $TPT_1 = 517{,}342$ person-weeks and $TPT_0 = 627{,}732$ person-weeks. The infection incidence rate and the infection incidence rate ratio are

$$\begin{aligned} IR_1 &= \frac{411}{517{,}342} \\ &= 0.794 \text{ cases per } 1{,}000 \text{ person} - \text{weeks}, \end{aligned}$$

$$\begin{aligned} IR_0 &= \frac{759}{627{,}732} \\ &= 1.210 \text{ cases per } 1{,}000 \text{ person} - \text{weeks}, \end{aligned}$$

and $IRR = 0.657$. All three estimates are close or equal to the true value (0.80, 1.20 and 0.667, see Sect. 7.9).

One statistical test to compare person-time data is a conditional exact test, based on the binomial distribution of the number of infected cases in the vaccinated group, given the total number of cases and the total person-times at risk. Under the Poisson assumption, c_1 is binomially $\mathrm{B}(c, p)$ distributed, conditional on $c = c_1 + c_0$, the total number of cases, and

$$p = \frac{\lambda_1 TPT_1}{\lambda_1 TPT_1 + \lambda_0 TPT_0}.$$

The null hypothesis that the two forces of infection are equal, $H_0 : \lambda_1 = \lambda_0$, is tested by testing the equivalent null hypothesis

$$H_0 : p = \frac{TPT_1}{TPT_1 + TPT_0}.$$

The null hypothesis can be tested with the SAS function PROBBNML(q, n, m), which returns the probability that an observation from a binomial distribution, with probability of success q and number of studies n, is less than or equal to m. If $r = TPT_1/TPT_0$, the ratio of the total person-times, then

$$\phi = \frac{p}{4r(1-p)}.$$

Exact binomial confidence limits for p are translated to exact confidence limits for $\phi = \lambda_1/\lambda_0$, the relative force of infection. If LCL_p and UCL_p are the exact lower and upper 100(1−α)% confidence limits for p (see Sect. 3.5.1), then

$$LCL_\phi = \frac{LCL_p}{r(1-LCL_p)},$$

and

$$UCL_\phi = \frac{UCL_p}{r(1-UCL_p)}$$

are exact 100(1−α)% confidence limits for ϕ.

Example 8.1 (continued) To test the null hypothesis that the relative force of infection equals 1.0, the null hypothesis

$$H_0 : p = \frac{517{,}342}{1{,}144{,}780} = 0.452$$

must be tested. Under the null hypothesis, the probability that the number of cases in the vaccine group is less or equal to 400 given that the total number of cases is 1,154

$$\text{PROBBNML}(0.452,\ 1170,\ 411) < 0.0001.$$

The exact 95% two-sided confidence interval for p is

$$(0.324, 0.379).$$

With $r = 517,342/627,732 = 0.824$, the following lower and upper limit of the 95% confidence interval for the relative cumulative force of infection ϕ are obtained as

$$LCL_\phi = \frac{0.324}{0.824(1-0.324)} = 0.582$$

and

$$UCL_\phi = \frac{0.379}{0.824(1-0.379)} = 0.741.$$

The interval contains the true value of ϕ, 0.667 (see Eq. (7.7)).

The incidence rate *IR* is an unbiased estimator only when the forces of infection are homogeneous. A class of infectious diseases with non-homogeneous forces of infection is seasonal infectious diseases; instead, their force of infection functions is cyclic. Seasonal change in the occurrence of infectious diseases is a common phenomenon in both temperate and tropical climates. Examples of seasonal infectious diseases are influenza and malaria, the childhood diseases measles, diphtheria and varicella, and the faecal–oral infections cholera and rotavirus.

8.2.2 Poisson Regression

An alternative to the analysis in the previous section is Poisson regression. When the forces of infection are homogeneous, the SAS code for an analysis without covariates is

SAS Code 8.2 Poisson regression for comparing person-time data

```
proc genmod data=Cohorts;
   class Group;
   model Case=Group / dist=Poisson offset=Logtime;
   estimate "IR1" intercept 1 GROUP  0 1;
   estimate "IR0" intercept 1 GROUP  1 0;
   estimate "IRR" intercept 0 GROUP -1 1;
   where Version="open" & FOI="homogeneous";
run;
```

`Group`=1 for the vaccinated subjects, and `Group`=0 for the unvaccinated subjects. The variable `Logtime` must contain the log-transformed time to infection (if `Case`=1) or the log-transformed length of the surveillance period/duration of the membership for non-infected subjects/non-infected subjects leaving the cohort before the end of the surveillance period (if `Case`=0).

SAS Output 8.2A Open cohort, homogeneous forces of infection

```
Label      Mean              Mean
         Estimate      Confidence Limits
IR1        0.0008
IR0        0.0012
IRR        0.6570      0.5827      0.7408
```

Example 8.1 (continued) The returned values for $IR_0, IR_1, IRR, LCL_\phi$ and UCL_ϕ are the same or almost the same values as found in the previous section. The advantage of Poisson regression over the exact conditional analysis is its simplicity, and that it allows the inclusion of covariates in the statistical model.

When it cannot be assumed that the forces of infection are homogeneous, in case of a seasonal infection, for example, the person-times must be stratified by follow-up time. Assume that a member of the cohort has a person-time of 43 weeks, and that it concerns a case, and that a second member has a person-time of 78 weeks and that this observation is censored. If the surveillance period is divided into strata of, say, 20 weeks, then the stratified data for these two members look like this

```
Member    Stratum  Time  Case     Member    Stratum  Time  Case
     A  stratum 1   20     0           B  stratum 1   20     0
     A  stratum 2   20     0           B  stratum 2   20     0
     A  stratum 3    3     1           B  stratum 3   20     0
                                       B  stratum 4   18     0
```

The CLASS and MODEL statements in SAS Code 8.2 must be changed to

```
class Group Stratum;
model Case=Group Stratum / dist=Poisson offset=Logtime;
```

When the new code is run, the output that is produced is

SAS Output 8.2B Open cohort, homogeneous forces of infection, stratified analysis

```
Label        Mean              Mean
         Estimate     Confidence Limits
IR1        0.0001
IR0        0.0002
IRR        0.6440     0.5725      0.7210
```

When the above SAS code is run with

```
where version="open" & FOI="non-homogeneous";
```

the following output is obtained:

SAS Output 8.2C Open cohort, non-homogeneous forces of infection, stratified analysis

```
Label      Mean           Mean
         Estimate    Confidence Limits
IR1       0.0001
IR0       0.0003
IRR       0.4875     0.4412      0.5386
```

Here also the confidence interval contains the true value of ϕ, 0.500 (see Eq. (7.9)). Note that with the stratified analysis the values for the estimates IR1 and IR0 are no longer meaningful.

8.2.3 Cox Regression

Person-times are time-to-event data, which means that the tools of survival analysis can be applied to analyse person-time data of a cohort, for example, Cox regression, using PROC PHREG. For simple analyses, only the MODEL statement is required, see SAS Code 8.3. In the MODEL statement, the value for censored observations (here: 0) must be specified. The parameter being estimated is the ratio of the force of infection functions ϕ, which in survival analysis is called the hazard ratio (see Appendix F). If the RISKLIMITS option is specified, 95% confidence limits for ϕ are returned.

SAS Code 8.3 Cox regression for comparing two sets of infection incidence data

```
proc phreg data=Cohorts;
   class Group (ref="0");
   model Time*Case(0)=Group / risklimits;
   where Version="open" & FOI="non-homogeneous";
run;
```

When the data are analysed with SAS Code 8.3, the following output is returned:

SAS Output 8.3 Open cohort, homogeneous forces of infection

```
Hazard      95% Hazard Ratio
 Ratio     Confidence Limits
 0.501     0.451        0.558
```

Cox regression requires the proportional hazards assumption. This is the assumption that the ratio $\lambda_1(t)/\lambda_0(t)$ does not depend on t, but is a constant (ϕ). The assumption can be tested by adding

```
assess ph resample;
```

When the force of infection of an infectious disease is non-homogeneous, in case of a seasonal infectious disease, for example, the proportional hazards assumption is perfectly reasonable. It is the assumption that the proportion of cases prevented by the vaccine is independent of the force of infection, i.e. the same whether the force of infection is high or low.

8.2.4 Comparing Two Nelson–Aalen Risk of Infection Estimates

The Nelson–Aalen estimator is a non-parametric estimator of the cumulative force of infection [53–56]. It does not require that the forces of infection are homogenous or proportional. The estimator is given by

$$\hat{\Lambda}_{NA} = \Sigma_{TPT_i} \frac{d_i}{r_i},$$

where d_i is the number of cases at TPT_i and r_i the total number of subjects at risk at TPT_i.

The Nelson–Aalen estimator is available in PROC LIFETEST, see SAS Code 8.4.

SAS Code 8.4 Nelson–Aalen estimate of cumulative force of infection

```
proc lifetest data=Cohorts nelson;
    time Time*Case(0);
    strata Group;
    where Version="open" & FOI="non-homogeneous";
run;
```

Example 8.1 (continued) The Nelson–Aalen estimates are $\hat{\Lambda}_{NA1} = 0.130$ and $\hat{\Lambda}_{NA0} = 0.259$. These values are (almost) identical to the true values, 0.130 and 0.260 (see Sect. 7.9).

To compare two Nelson–Aalen cumulative force of infection estimates, several parametric tests exist [57]. Here, a non-parametric test based on bootstrapping is presented. The advantage of this approach is that it does not only allow to compare two Nelson–Aalen cumulative force of infection estimates but also to Nelson–Aalen risks of infection estimates. The Nelson–Aalen estimate of the risk of infection is obtained by applying the exponential formula in Sect. 7.8.2:

$$\hat{\pi}_{NA} = 1 - \exp(-\hat{\Lambda}_{NA}).$$

Example 8.1 (continued) The Nelson–Aalen estimate of π_1 is

$$\hat{\pi}_{NA1} = 1 - \exp(-0.130) = 0.228,$$

and

$$\hat{\pi}_{NA0} = 1 - \exp(-0.259) = 0.122.$$

The Nelson–Aalen estimate of the relative risk of infection is

$$\hat{\theta}_{NA} = \frac{0.122}{0.228} = 0.535.$$

These values too are almost identical to the true values 0.229, 0.122 and 0.533, see Sect. 7.9.

To compare two Nelson–Aalen risks of infection estimates by means of bootstrapping, a very large number N of random samples with replacement are drawn from the original data set, the bootstrap samples. Each bootstrap sample must be of the same size of the original data set. For each bootstrap sample, the derived attack rates $\hat{\pi}_{bs\,1}$ and $\hat{\pi}_{bs\,0}$, as well as the bootstrap estimate for the relative risk of infection

$$\hat{\theta}_{bs} = \frac{\hat{\pi}_{bs\,1}}{\hat{\pi}_{bs\,0}}.$$

The N bootstrap estimates form an empirical sampling distribution for

$$\hat{\theta}_{NA} = \frac{\hat{\pi}_{NA1}}{\hat{\pi}_{NA0}}.$$

A confidence interval for the relative risk of infection θ can be obtained by the so-called percentile method [58]. In this method, the lower and upper limit of the 95% confidence interval are the 2.5th and the 97.5th percentile of the empirical sampling distribution.

Example 8.1 (continued) A bootstrap 95% confidence interval for the relative risk of infection θ can be obtained with the SAS Code I.3 in Appendix I. When this code is run, as confidence interval for θ the interval

$$(0.484, 0.589)$$

is found. Note that this interval is similar to the interval found in Sect. 8.1.

A bootstrap confidence interval will not be reproducible when N, the number of bootstrap samples, is too low. This can be checked by running the SAS code twice, with the number of bootstrap samples fixed. For example, when N was set to 10,000 (see REPS= option), the second interval returned was

$$(0.485, 0.591).$$

Thus, with 10,000 bootstrap samples the interval is reproducible.

8.3 Recurrent Infections

A *recurrent infection* is an infection caused by the same pathogen in a subject who has experienced at least one infection before. The infection is called a *reinfection* when the pathogen that caused the previous infection was eradicated by treatment. When the pathogen is not completely eradicated and the infection returns, the infection is called *persistent*. A notorious example of a persistent infection is chronic urinary tract infection or UTI. A UTI is the result of a bacterial infection. The bacteria enter the urinary system through the urethra and then multiply in the bladder. The most common cause is the bacteria *Escherichia coli* (*E. coli*).

The statistical analysis of recurrent infection data is complex. There are two major methodological challenges to address. The first challenge is the possibility that the risk of a next infection is affected by previous infections. It could, for example, be the case that a first infection makes a subject more susceptible for infection. The second challenge is that the risk of infection may be different among subjects. In that case, it may be that the more prone subjects are contributing more infections than less prone subjects. For neither challenge, a simple statistical solution exists. But there are other challenges as well. An example of an infectious disease with recurrent episodes in patients is malaria. A group of experts on this disease reported that it is difficult to define when a malaria episode has ended and when following treatment a subject becomes susceptible to a further episode [59]. For this reason and the statistical complexity involved in analysing recurrent episodes, the consensus of the group was that the primary endpoint in a malaria study should be time to first episode of clinical malaria, although the members admitted that the total number of episodes of malaria in study participants might better measure the total burden of malaria in the community. This is also pointed out by Janh-Eimermacher et al., who give an excellent overview of the statistical methods used to assess vaccine efficacy for the prevention of acute otitis media (AOM) by pneumococcal vaccination [60]. They note that: 'The proportion of subjects with at least one episode might be equal in two vaccine groups, whereas the groups differ substantially in the total number of episodes. (...) Decreasing the total number of AOM might reduce the total costs for treating AOM, improve the problem of antibiotic resistance caused by broad antibiotic use in treatment of AOM and reduce the long-term effects cause by recurrent episodes'. The drawback of considering only the first infection is that not all available information is used.

8.3.1 Average Number of Episodes Experienced by a Subject

The statistical analysis of intra-individual numbers of episodes depends on which of the following two conditions are met:

1. The risk of a next episode is unaffected by previous episodes.
2. The risk of infection does not differ between subjects.

If both conditions are met, the statistical analysis is straightforward. In that case, the infection incidence rates can be estimated total number of infections divided by the total person-time. The incidence rate ratio can be compared between groups by means of SAS Code 8.2.

Example 8.3 Fleming and Harrington present the data[1] of a randomized clinical study in patients with chronic granulomatous disease (CGD) [61]. The disease is an inherited (genetic) immune system disorder that occurs when the phagocytes dysfunction. As a result, the body is not protected from recurrent bacterial and fungal infections. Most people are diagnosed with CGD during childhood. In total, 128 patients were enrolled in the clinical study, of whom 65 were treated with placebo and 63 with interferon gamma. In the placebo group, the follow-up time varied between 91 and 439 days, the number of recurrent infections between 0 and 7; the total number of infections was 56, the total person-time 18,524 days, which corresponds to an average infection incidence rate of 0.0030 infections per person-day. In the interferon gamma group, the follow-up time varied between 160 and 414 days and the number of recurrent infections between 0 and 3. In this group, the total number of infections was 20 and the total person-time 18,953 days. This corresponds to an average infection incidence rate of 0.0011 infections per person-day. Infection incidence rate ratio was thus

$$IRR = \frac{20/18,953}{56/18,524} = 0.349.$$

As said, the incidence rates can be compared using SAS Code 8.2. Alternatively, SAS Code 8.5 can be used, in which intra-individual numbers of infections are compared between the two treatment groups. The two analyses yield the same results. The

[1]Online available on http://www.mayo.edu/research/documents/cgdhtml/doc-10026922.

variable `Logfutime` must contain the log-transformed follow-up times (variable `Futime`).

SAS Code 8.5 Analysis of numbers of recurrent infections [61]

```
proc genmod data=GCD;
   class Rx;
   model Nevent=Rx / dist=Poisson offset=Logfutime;
   estimate "placebo"                          Intercept 1 rx  0 1 / exp;
   estimate "Interferon gamma"                 Intercept 1 rx  1 0 / exp;
   estimate "Infection Incidence Rate Ratio"  Intercept 0 rx -1 1 / exp;
run;
```

SAS Output 8.5

```
Label                                   Estimate    Confidence Limits
Exp(placebo)                             0.0011      0.0002    0.05
Exp(Interferon gamma)                    0.0030      0.0023    0.0039
Exp(Infection Incidence Rate Ratio)      0.3491      0.2095    0.5816
```

In practice, it will be unlikely that the second condition is met. More likely is that the risk of infection differs between subjects. This is called overdispersion. If overdispersion is ignored, the variance of the estimator is underestimated. Let the random variable **Y** be the intra-individual number of episodes. If the Poisson model holds, then the variance of **Y** will be equal to its expectation:

$$\mathrm{var}(\mathbf{Y}) = \mathrm{E}(\mathbf{Y}) = \lambda.$$

A much applied approach to deal with count data that exhibit overdispersion is to assume that the individual risk itself may be regarded as a random variable. In that case, the probability distribution will be a compound (mixed) distribution with

$$\mathrm{var}(\mathbf{Y}) = V(\lambda) > \lambda.$$

The standard choice for the variance function is

$$V(\lambda) = \sqrt{\zeta}\lambda, \quad \zeta > 1.$$

When $\zeta > 1$, the Poisson model is overdispersed. When this model is fitted, the parameter estimates do not changes, but the estimated covariance matrix is inflated. The dispersion parameter ζ can be estimated by specifying the DSCALE option in the MODEL statement. When this code is run, first, estimates are obtained for the non-overdispersed Poisson model. Next, a scale parameter ($\sqrt{\zeta}$) is estimated, and the standard errors are adjusted by multiplying them by the value for the scale statistic,

1.1311, making the statistical tests more conservative. The confidence interval for *IRR* based on the overdispersed Poisson model is

$$(0.196\,,\ 0.622),$$

a slightly wider interval than that based on the non-overdispersed model (see SAS Output 8.5).

8.4 Sample Size Estimation

8.4.1 Studies Comparing Two Attack Rates

To estimate the required sample size of a vaccine field efficacy study comparing two attacks rates, the parameters to be specified are the expected attack rate in the control group and the expected vaccine efficacy. With the vaccine efficacy specified, the expected attack rate in the investigational vaccine group can be calculated. The required sample size can then be calculated with SAS Code 3.2.

If super efficacy must be demonstrated, more complicated calculations are required. The formula is given by Farrington and Manning [62]. For the SAS code, see SAS Code I.4 in Appendix I.

Example 8.4 Consider a placebo-controlled influenza vaccine field efficacy study. Assume that the expected attack rate in the control group is 0.25 and that the expected vaccine efficacy is 0.8. If the null hypothesis is that the vaccine efficacy is zero, then SAS Code 3.2 calculates the sample size to be $2 \times 65 = 130$ to be secured of a power of 0.9 (for a two-sided significance level of 0.05). The FDA/CBER criterion for super efficacy for influenza vaccines is 0.4. When the SAS Code I.4 is run, a sample size of $2 \times 147 = 294$ is found.

8.4.2 Studies Comparing Two Infection Incidence Rates

The procedure to determine the sample size for a vaccine field efficacy study in which two infection rates are to be compared is as follows. Let c_0 denote the expected number of placebo cases, c_1 the expected number of cases in the investigational vaccine group and $c = (c_0 + c_1)$ the expected total number of cases. For c fixed, determine the largest number for c_1 such that the lower limit of the two-sided $100(1-\alpha)\%$ confidence interval for the vaccine efficacy ϑ_ϕ is > 0, or the criterion if super efficacy is to be demonstrated. The actual power of the study is then the probability that the number of cases in the investigational vaccine group is less or equal to c_1. Find the value for c such that the actual power is equal to the desired power.

The formula for the expected total number of cases is

$$\begin{aligned} c &= n_0\lambda_0 + n_1\lambda_1 = n_0\lambda_0 + n_1\lambda_0(1-\vartheta_\phi) \\ &= n_0\lambda_0 + n_0 r\lambda_0(1-\vartheta_\phi) = n_0\lambda_0[1+r(1-\vartheta_\phi)], \end{aligned}$$

where λ_0 and λ_1 are the expected infection rates during the length of the planned average surveillance period, and $r = n_1/n_0$, the randomization ratio. Hence,

$$n_0 = \frac{c}{\lambda_0[1+r(1-\vartheta_\phi)]}.$$

Example 8.5 An investigator wants to estimate the sample size for a vaccine field efficacy study with a planned average surveillance period of 24 months, a 2:1 randomization ratio, an expected infection rate in the placebo group of 0.05 cases per 24 months, an expected vaccine efficacy of 0.7 and with 0.4 as bound for super efficacy. The sample size can be obtained with SAS Code I.5 in Appendix I. If the desired power is 0.9, then the required number of investigational cases is 40 and the required number of total cases 92. This corresponds to a sample size of 3,453 subjects, 2,302 to be randomized to the investigational vaccine and 1,151 to placebo.

8.4.3 Studies Comparing Two Forces of Infection

Power and sample size calculations for comparing two forces of infection can be performed with the TWOSAMPLESURVIVAL statement of PROC POWER. If a proportional hazards model is assumed, the TEST= option should be set to TEST=LOGRANK. Sample size estimation for comparing two forces of infection requires detailed specification of the design of the study. The reader is therefore referred to the chapter on PROC POWER in the SAS/STAT User's Guide for a description of the different options to specify the details of the design of the study.

Chapter 9
Vaccine Effectiveness Studies

Abstract Vaccine effectiveness studies are observational studies, and thus at risk of confounder bias. This chapter therefore opens with a discussion of what confounding is and how it can be eliminated. Many vaccine effectiveness studies have a so-called case-referent design. The key idea of these designs is clarified, and the major case-referent designs—including the popular test-negative design—are inspected. The important but difficult to grasp difference between cumulative reference sampling and incidence sampling is explained. Next, adjusted data analysis to control for confounding, either by means of stratification or regression, is discussed. The chapter concludes with considerations on how to represent and select confounders in a regression model.

9.1 Confounding

9.1.1 Definition

Vaccine effectiveness studies are observational studies, and like all observational studies they have this Achilles heel: the risk of confounder bias.

Confounding is a mixing of effects which distorts the relationship between vaccination and infection, and therefore needs to be eliminated. A variable is a confounder of the effect of vaccination on the risk infection when it is causally related with both getting vaccinated and becoming infected. Here 'A is a cause of B' should be interpreted in the none-strict sense that there is a statistical dependence directed from A to B. Thus, when the susceptibility to infection increases with age, then it is said that age is a 'cause' of infection.

Confounding can be graphically displayed using a directed acyclic graph (DAG), see Fig. 9.1. The arrows represent causative relations directed from C (confounder) to V (vaccination) and from C to I (infection) [63, 64].

Example 9.1 Assume that compared to individuals living in a rural area, individuals living in urban area have (*i*) easier access to health care and thus to getting vaccinated, but (*ii*) are at a higher risk of being infected. Then living area is a confounder of the

J. Nauta, *Statistics in Clinical and Observational Vaccine Studies*,
Springer Series in Pharmaceutical Statistics,
https://doi.org/10.1007/978-3-030-37693-2_9

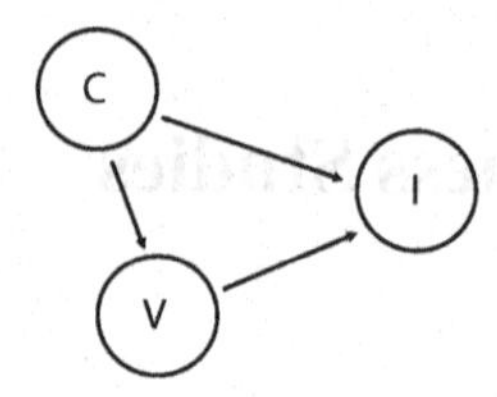

Fig. 9.1 DAG illustrating confounding (C = Confounder V = Vaccination I = Infection)

Table 9.1 Hypothetical vaccine effectiveness data to illustrate confounding

Infected	Urban area		Infected	Rural area	
	Vaccinated			Vaccinated	
	Yes	No		Yes	No
Yes	10	200	Yes	3	135
No	990	3,800	No	597	5,265
Total	1,000	4,000	Total	600	5,400

effect of vaccination on infection. To see this, consider the (hypothetical) data in Table 9.1. For a vaccinated urban person, the relative risk of becoming infected is

$$\frac{10/1{,}000}{200/4{,}000} = 0.2.$$

For a vaccinated rural person, the relative risk of becoming infected is

$$\frac{3/600}{135/5{,}400} = 0.2.$$

When the data are pooled, i.e. when living area is ignored, the relative risk of infection for vaccinated persons is found to be

$$\frac{13/1{,}600}{335/9{,}400} = 0.228.$$

This difference, this bias, is due to living area being a confounder.

Denote the stratum-specific relative risks by θ_s (= 0.2) and the pooled relative risk by θ_p (= 0.228). Let π_u be the risk of infection for an unvaccinated urban person, π_r the risk of infection for an unvaccinated rural person, f_v the fraction of vaccinated persons living in a city and f_u the fraction of unvaccinated persons living in a city. With a little algebra it can be shown that

$$\theta_p = \frac{[f_v\pi_u + (1 - f_v)\pi_r]\theta_s}{f_u\pi_u + (1 - f_u)\pi_r}.$$

Sufficient conditions for $\theta_p = \theta_s$ are $\pi_u = \pi_r$ or $f_v = f_u$. If $\pi_u = \pi_r$, then living area is not a cause of infection. If $f_v = f_u$, then living area is not a cause of vaccination. Thus, for living area to be a confounder, it must be a cause of getting vaccinated and of becoming infected. In the example, $f_v = 0.625$ and $f_u = 0.426$. In the numerator, more weight is given to vaccinated urban persons, while in the denominator more weight is given to unvaccinated rural persons. In other words, the positive effect of vaccination is confounded by (mixed-up with) the negative effect of living in a city, with as net result overestimation of the relative risk of infection, and thus underestimation of the vaccine effectiveness.

The check for the presence of confounding is to compare the vaccination rates between the two strata:

$$\frac{1{,}000}{5{,}000} = 0.2$$

versus

$$\frac{600}{6{,}000} = 0.1$$

and next, to compare the risk of infection for unvaccinated subjects between the two strata:

$$\frac{200}{4{,}000} = 0.05$$

versus

$$\frac{135}{5{,}400} = 0.025.$$

Because both the vaccination rate and the risk of infection for unvaccinated persons differ between the two strata, living area is a confounder.

9.1.2 Eliminating Confounding

There are two approaches to eliminating confounding. Confounding can be eliminated by the design of the study or in the statistical analysis stage.

Controlling for confounding in the design phase of the study can be done by either restricting or matching. The first method, a somewhat unsatisfactory method, is to avoid confounding by making the source population more homogeneous. In Example 9.1 this would mean to restrict the population to either persons living in a city or to persons living in the country side. Unsatisfactory indeed, because by making the source population more homogenous, the generalizability of the study result is reduced. The second method is matching on confounders. The aim of matching is to make the vaccinated and the unvaccinated subjects as similar as possible with regard to the confounders. For every vaccinated subject, one or more unvaccinated subjects are matched according to, say, age and sex. Matching has some attractive advantages, such as a potential increase in statistical power for example. The big disadvantage of the method is that when multiple confounders are to be controlled for, it may be difficult if not impossible to find matching unvaccinated subjects. For these reasons, matching is not discussed here, with the exception of time-matching of cases and controls in nested case-control studies (see Sect. 9.2.2).

In the statistical analysis stage, confounding can be adjusted for (controlled for) by stratification, regression or the propensity score method. The stratification should be done in such a way that within the strata there is no confounding. In Example 9.1, confounding can be adjusted for by stratifying on living area. For each separate stratum, an effectiveness estimate is obtained, and these stratum-specific estimates are then averaged, if homogeneous. The resulting estimate is then confounder-adjusted. Regression adjustment is achieved by including the confounders as covariates in the regression model. The propensity score method removes confounding by balancing baseline covariate values between vaccinated and unvaccinated subjects.

9.1.3 Some Known or Potential Confounders

Age is almost certainly a confounder, because both vaccine uptake and risk of infection tend to vary by age. Young children are more susceptible to infection due to the immaturity of their immune system. Elderly are more susceptible to infection due to *immunosenescence*, the gradual, age-related deterioration of the immune system.

Women are usually at a higher risk of infection through greater contact with children. Thus, when sex is related to health care utilization and vaccination, sex will be a confounder.

Health care seeking behaviour, i.e. a person's propensity to seek care when ill, is a likely confounder because such behaviour may stimulate vaccine uptake and reduce the risk of infection due to behaviours like frequent handwashing.

Another likely confounder is high-risk status, i.e. presence of conditions that increases a person's risk of infection or increase the risk of complications related to the infection. High-risk status may therefore be a confounder when it increases a person's likelihood of being vaccinated.

A very common type of confounding in observational studies is confounding by indication. Persons who have an increased risk of infection, health care workers, for example, will get vaccinated because of this risk. And because of the increased risk of infection, indication is a confounder. The net result is called negative confounding, because the association being vaccination and infection is biased towards null. The alternative is called healthy vaccinee bias and refers to a situation when persons who are in better health conditions are more likely to adhere to vaccination recommendations. Now the net result will be positive confounding because the association being vaccination and infection will be biased towards one.

Functional and cognitive limitations have also been shown to be important confounders.

9.1.4 *Confounding Versus Effect Modification*

When the vaccine effectiveness varies by sex, then it is said that sex is a vaccine effect modifier (effect modifier for short). Effect modification occurs when the magnitude of the effect of vaccination on the occurrence of infection differs depending on the level of a third variable, the effect modifier. In this situation, computing an overall vaccine effectiveness estimate would be misleading. The difference between confounding and effect modification is that confounding must be eliminated to avoid bias, whereas effect modification should not be eliminated but studied.

Effect modification is similar to interaction but not the same as. Interaction refers to the effects of two interventions while effect modification refers to the effect of one intervention varying across levels of another, non-interventional, variable [65]. A variable can be both a confounder and an effect modifier.

Example 9.1 (continued) Assume that the number of vaccinated urban persons who got infected was 15 instead of 10. Then the relative risk of infection for a vaccinated urban person would be 0.3, which is a higher relative risk than that for vaccinated rural person (0.2). Thus, living area is an effect modifier. But it is also a confounder, because the results of the quick check do not change. When the data are pooled, the relative risk of infection for vaccinated is

$$\frac{18/1{,}600}{335/9{,}400} = 0.316,$$

a value which is obviously meaningless.

Despite the difference between the two concepts, in the statistical analysis effect modification can be tested in the same manner as interaction is tested, by adding the appropriate product term to the model. If the term is statistically significant, the relative risk of infection should be reported for each stratum separately.

9.2 Case-Referent Designs

9.2.1 *Key Idea of Case-Referent Designs*

Most vaccine effectiveness studies have a so-called case-referent design. In this design, all the cases occurring in the source population are enrolled and, separately, a random sample from the source population or from its non-case segment. Why many vaccine effectiveness studies have a case-referent design is easiest explained by making the comparison with effectiveness studies with a cohort design.

Vaccine effectiveness studies with a cohort design, in which the vaccination status is determined for *all* members of the source population, are rare. To determine the vaccination status of all members is often not doable, due to the absence of an immunization registry for example. Also, when confounder data must be collected the costs may be prohibitive, when expensive clinical tests are required. Consider a cohort study with c_1 vaccinated cases, c_0 unvaccinated cases, n_1 vaccinated members and n_0 unvaccinated members. Then the standard error of the log-transformed rate ratio is

$$SE(\log RR) = \sqrt{1/c_1 - 1/n_1 + 1/c_0 - 1/n_0},$$

(see Sect. 3.5.2) The precision of $\log(RR)$ is determined by the number of cases and less by the sizes of the sub-cohorts, because for not too small sizes $1/n_0$ and $1/n_1$ are approximately zero. Now,

$$RR = \frac{c_1/n_1}{c_0/n_0},$$

which can be rewritten as

$$RR = \frac{c_1/c_0}{n_1/n_0},$$

the ratio of the cases divided by the ratio of the sizes of the subgroups. To be able to calculate the ratio n_1/n_0, the vaccination status of all members of the cohort must be known. But this ratio can also be estimated, from a random sample from the cohort. RR then becomes

$$RR = \frac{c_1/c_0}{s_1/s_0},$$

where s_1 the number of vaccinated members in the random sample and s_0 the number of unvaccinated members. Thus, instead of having to determine the vaccination status for all members of the cohort, it is sufficient to do this for a random sample only, without (too much) loss of precision. *This is the key idea of case-referent designs.* In a case-referent design, the vaccination status of the cases of the infectious disease in the source population is compared with that of a sample from the members of the source population, the referents. The referents should represent the source population in terms of vaccination status. The big advantage of case-referent designs is, as explained, that data needs to be collected on only a subset of the source population.

Example 9.2 Consider a cohort study with $c_1 = 10$, $c_0 = 80$, $n_1 = 1{,}000$ and $n_0 = 4{,}000$. Then $RR = 0.500$, and

$$SE(\log RR) = \sqrt{1/10 - 1/1{,}000 + 1/80 - 1/4{,}000} = 0.334.$$

Assume that a random sample from 500 referents is drawn from the cohort, $s_1 = 94$ vaccinated and $s_0 = 406$ unvaccinated. Then $RR = 0.540$, and, if there is no overlap between the cases and the referents (see Sect. 9.4.4)

$$SE(\log RR) = \sqrt{1/c_1 + 1/s_1 + 1/c_0 + 1/s_0}$$
$$= \sqrt{1/10 + 1/94 + 1/80 + 1/406} = 0.354.$$

Thus, determining the vaccination status for only a random sample from the source population means a minor loss in precision, but usually a large saving in costs.

9.2.2 *Major Case-Referent Designs*

When the source population is a cohort, there are basically two case-referent designs, the case-cohort design and the case-control design. The two designs differ with respect to the set of individuals from which the referents are sampled and how they are sampled.

In the case-cohort design, the referents are sampled from the whole cohort, either at the start or at the end of the surveillance period. An already sampled referent may become a case during the surveillance period, or a case can afterwards be sampled as a referent. In other words, there can be overlap between the cases and the referents, and in the statistical analysis this overlap should not be ignored (see Sect. 9.4.4). The parameter being estimated in a case-cohort design is the same as that being estimated in a cohort design, the relative risk of infection θ for the members of the cohort.

In the case-control design, there are two different approaches to how the referents (the 'controls') are sampled. In the first approach, the controls are sampled at the end of the surveillance period, not from the whole cohort as in the case-cohort design, but from the non-cases, i.e. the subgroup of cohort members still infection-free at the end of the surveillance period. This type of control sampling is called cumulative sampling. The controls represent cohort membership. The parameter that can be estimated is the relative risk of infection θ, provided that the so-called 'rare disease assumption' is met. This is the assumption that the risk of infection is low. The assumption is required because the ratio vaccinated/unvaccinated in the cohort is estimated by the ratio vaccinated/unvaccinated among the controls. When the cohort is open, both the case-cohort and the case-control design with cumulative sampling require the assumption that the reason for leaving is independent of the vaccination status.

When the source population is a cohort, the case-control design is often called a 'nested' case-control design. The nested case-control design with cumulative sampling is very similar to the case-cohort design. An important difference is that the case-cohort design does not require a low infection risk. This raises the question why many investigators prefer the nested case-control design with cumulative sampling to the case-cohort design. First, for most infectious diseases, the risk of infection is indeed low, and second and perhaps more important, the statistical analysis of case-control data is much simpler than the analysis of case-cohort data. Especially regression analysis of case-cohort data is complex.

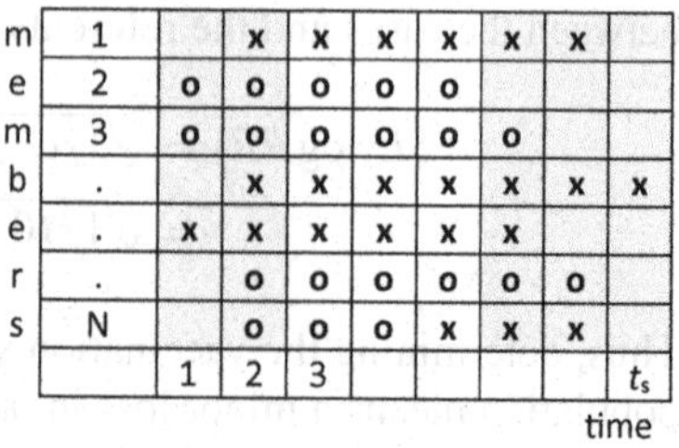

Fig. 9.2 Graphical illustration of risk set membership over time

m	1		x	x	x	x	x	x	
e	2	o	o	o	o	o			
m	3	o	o	o	o	o	o		
b	.		x	x	x	x	x	x	x
e	.	x	x	x	x	x	x		
r	.		o	o	o	o	o	o	
s	N		o	o	o	x	x	x	
		1	2	3					t_s

time

The second approach to control sampling is sampling during the surveillance period. This is called density sampling (also incidence density sampling). Controls are sampled from the risk set (i.e. the set of cohort members still infection-free at the time of sampling) either every time that a case occurs, or throughout the surveillance period, independent of the occurrence of the cases. When the controls are sampled every time that a case occurs, it is said that the controls are time-matched to the cases. Usually 1–5 controls per case are sampled, because the sampling of more controls improves the statistical power only minorly. This type of sampling is called density sampling because the controls now represent person-time rather than cohort membership. To see why, consider a design with density sampling throughout the surveillance period, at random time points. In Fig. 9.2 a matrix is displayed, illustrating risk set membership over time. The rows represent members and the columns represent time. If in cell(i, j) there is an **x**, then the ith subject was a vaccinated member of the risk set at time point j; if there is an **o**, then he was an unvaccinated member at the time point; if the cell is empty, the subject was not a member of the risk set at the time point. The sum of all the **x** is the total person-time TPT_1 for the vaccinated members, and the sum of all the **o** is the total person-time TPT_0 for the unvaccinated members. If controls are sampled at random throughout the surveillance period, then the probability that a vaccinated member is sampled is TPT_1/TPT, where

$$TPT = TPT_1 + TPT_0.$$

Likewise, the probability that an unvaccinated member is sampled is TPT_0/TPT. Thus, if s controls are sampled then the expected values of the total number of vaccinated controls s_1 and the total number of unvaccinated controls s_0 are

$$\mathrm{E}(\mathbf{s}_1) = \frac{s \times TPT_1}{TPT},$$

and

$$\mathrm{E}(\mathbf{s}_0) = \frac{s \times TPT_0}{TPT}.$$

Hence, the ratio s_1/s_0 estimates the ratio of the total person-times.

When the controls are not sampled at random time points, but, for example, every time a case occurs, then the ratio s_1/s_0 will estimate the ratio of the total person-times at the time of sampling rather than the ratio of the total person-times across the whole surveillance period. In that case, a stratified analysis is required, with the strata being defined by the cases and the time-matched controls.

The parameter that can be estimated in case of density sampling is the relative force of infection ϕ. Density sampling has two advantages over cumulative sampling. First, it does not require that the ratio vaccinated/unvaccinated in the cohort to be steady over time. Second, the rare disease assumption is not needed.

On the choice between time-matched and unmatched density sampling of controls, the following can be said [66]: assume that the relative force of infection ϕ is a constant, then time-matching is required if the ratio of vaccinated members to unvaccinated members varies over the surveillance period. There are several reasons why this ratio can vary over time. If vaccine supplies are limited, for example, in case of a pandemic, some members of the cohort will only be vaccinated after the start of the surveillance period and then the ratio will vary over time. In case of time-matched sampling, the statistical analysis must be stratified by calendar time, with each stratum containing a case and its time-matched controls. If the ratio of vaccinated members to unvaccinated members does not vary over the surveillance period, time-matched sampling is not required.

In a nested case-control design with density sampling, a cohort member can be sampled more than once as a control, and a control can become a case later during the surveillance period. (Note that a case cannot become a control afterwards, because cases are no longer at risk.) So, here too there can be overlap between the cases and the controls, but also between the controls. Though it may be very counterintuitive, here this overlap need not be taken into account in the statistical analysis. If a cohort member is enrolled in the study more than once, his data must be entered repeatedly, as if from different members [67].

When the source population is a dynamic population, the only possible case-referent design is the case-control design with time-matched density sampling. The control sampling must be time-matched because the ratio of vaccinated members to unvaccinated members will usually not be a constant in a dynamic population. Here too, the parameter that can be estimated is the relative force of infection ϕ, without the need for the rare disease assumption. In a dynamic population, cumulative sampling is not possible.

9.2.3 The Test-Negative Design

The test-negative design (TND) is an epidemiologic design that claims to eliminate confounding due to health care seeking behaviour. The design has emerged in recent years as the preferred method for estimating influenza vaccine effectiveness. Patients seeking health care for an acute respiratory illness (ARI) are tested for influenza. Those who test positive are the cases and those who test negative are the referents.

The design eliminates confounding due to health care seeking behaviour by restricting the source population to those members who will seek health care for an ARI when infected. This, of course, reduces the generalizability of the results of a TND study. If the participating sites are hospitals, the enrolled subjects will usually be severe influenza cases. This further reduces the generalizability of the results of a TND study. TND vaccine effectiveness estimates are estimates of the effectiveness against (severe) medically attended influenza infection, and they should not be interpreted as estimates of the effectiveness against influenza infection in general.

The core assumption of the TND is that the occurrence of ARIs due to other respiratory pathogens than influenza does not differ between vaccinated and unvaccinated subjects [68].

It is sometimes argued that the TND is a special type of case-control design. Strictly speaking, this is not correct, because in a TND the referents are enrolled rather than sampled from the source population, as in a true case-control design. In Sect. 9.5.2, it will be argued that this has consequences for how the data of a TND study must be analysed.

As said, the TND eliminates confounding due to health care seeking behaviour by restricting the source population to the members who will seek health care for an ARI. This is a consequence of the applied case-finding strategy: enrolling not all influenza cases but only the cases who come to attention of health care workers. But this case-finding strategy is not unique to the TND, it is applied in as good as all observational influenza studies, because the more intensive strategy often applied in clinical influenza efficacy studies (testing all ARI cases for influenza) is simply not doable in effectiveness studies. In other words, not only the TND but almost all designs for observational influenza studies eliminate confounding due to health care seeking behaviour.

The TND design has also been applied in studies to estimate cholera vaccine effectiveness, rotavirus vaccine effectiveness and pneumococcal vaccine effectiveness.

9.3 The Odds Ratio and Its Prominent Role in Case-Referent Designs

The odds of an outcome is the ratio of the probability of the outcome occurring to the probability of the outcome not occurring. The odds that it will rain today, the betting odds that Flash will win the horse race, etc. An odds greater than one means that it is more likely that the outcome will occur than not occur. An odds ratio is, as implied by the name, the ratio of two odds. Epidemiologic research is occurrence research, and the tools epidemiologists use are occurrence measures and ratios of occurrence measures (effect measures). The odds of infection is not an occurrence measure, and thus not a useful epidemiologic measure, and the odds ratio is not a useful effect measure. Why then its prominent role in epidemiologic research? To understand this, the distinction between the odds ratio as a *parameter* and the odds ratio as an *estimator* must be made. As parameter, the odds ratio is, as noted, of

Table 9.2 Crude data analysis—statistical tests and corresponding estimators

Source population	Study design	Statistical test	Estimator	Effect measure
Cohort	Cohort	Logit test for a relative risk+	RR	θ
		Cox regression	IRR	ϕ
		Poisson regression		
	Case-cohort	Sato's chi-square test+	RR_{Sato}	θ
	Case-control with cumulative sampling	Logit test for an odds ratio+	OR	$\approx\theta$*
	With density sampling	Logit test for an odds ratio	OR	ϕ
Dynamic population	Case-control with density sampling	Logit test for an odds ratio	OR	ϕ

*If the rare disease assumption is met; +See Sects. 9.4.2–9.4.4

no special interest, but as estimator the odds ratio is of paramount importance. In case-control designs, the estimator is an odds ratio (Table 9.2):

$$OR = \frac{c_1/c_0}{s_1/s_0}.$$

Here c_1/c_0 is the odds of being vaccinated among the cases and s_1/s_0 is the odds of being vaccinated among the controls. But, what is being estimated is not an odds ratio (parameter), but either a relative risk of infection (nested case-control design with cumulative sampling) or a relative force of infection (case-control design with density sampling), i.e. the key effect measures in vaccine effectiveness studies.

9.4 Crude Data Analysis

9.4.1 Crude Vaccine Effectiveness

With *crude vaccine effectiveness* is meant the effectiveness estimate which is obtained if in the statistical analysis confounders and effect modifiers are ignored, and the only factors that are accounted for are vaccination status and infection. Sullivan and Cowling argue that the term crude vaccine effectiveness is a misnomer, and that crude vaccine effectiveness estimates indicate the correlation of vaccination with infection, but may not be an accurate estimate of the causal effect of vaccination on the risk of infection because that association may be confounded [69]. That argument is unanswerable. So, why then spend time on the crude analysis of vaccine effectiveness data? Because it will help in better understanding of the proper analysis, the confounder-adjusted analysis.

9.4.2 *Cohort Data*

The statistical methods in Sects. 8.1 and 8.2 for the analysis of vaccine efficacy data can also be applied for a crude analysis of vaccine effectiveness data from a cohort study.

Example 9.3 Khatib et al. report the results of an oral cholera vaccine effectiveness study in Zanzibar [43]. They offered two doses of a killed whole-cell B-subunit cholera vaccine to individuals aged two years or older in six rural and urban sites. The source population of the study was a cohort of 43,791 individuals eligible to receive the vaccine, of whom 23,921 received two doses. Between February 2009 and May 2010 there was an outbreak of cholera, which enabled the investigators to assess the vaccine effectiveness. Six of the 23,921 recipients of two vaccine doses had cholera episodes compared with 33 of the 20,500 unvaccinated people. Thus, the relative risk of infection for vaccinated individuals was

$$
\begin{aligned}
RR &= \frac{c_1/n_1}{c_0/n_0} \\
&= \frac{6/23{,}921}{33/20{,}500} = 0.156,
\end{aligned}
$$

with

$$
\begin{aligned}
SE(\log RR) &= \sqrt{1/c_1 - 1/n_1 + 1/c_0 - 1/n_0} \\
&= \sqrt{1/6 - 1/23{,}921 + 1/33 - 1/20{,}500} = 0.444.
\end{aligned}
$$

A lower and an upper 95% confidence limit for the relative risk of infection θ are

$$LCL_\theta = \exp[\log 0.156 - 1.96(0.444)] = 0.065$$

and

$$UCL_\theta = \exp[\log 0.156 + 1.96(0.444)] = 0.372.$$

Hence, the crude estimate of the oral cholera vaccine effectiveness ϑ is

$$VE = 1 - 0.156 = 0.844,$$

with

$$(0.628, 0.935)$$

as 95% confidence interval for ϑ. Despite the large cohort size, the confidence interval is rather wide, due to the low number of cases.

The above estimator and its standard error can be transformed into a z-test:

$$z = \frac{\log(RR)}{SE(\log RR)}.$$

This test is sometimes called the logit test for a relative risk.

9.4.3 Case-Control Data

The crude analysis of case-control data does not depend on how the controls were sampled, by means of cumulative sampling or by density sampling. Let s_1 be the number of vaccinated controls and s_0 the number of unvaccinated controls. The formula for the odds ratio is

$$OR = \frac{c_1/c_0}{s_1/s_0}.$$

A more common notation is the 'cross-product' notation:

$$OR = \frac{c_1 \times s_0}{s_1 \times c_0}.$$

The asymptotic standard error of $\log(OR)$ is usually estimated by

$$SE(\log OR) = \sqrt{1/c_1 + 1/s_1 + 1/c_0 + 1/s_0}. \tag{9.1}$$

What does depend on how the controls were sampled is the interpretation of the odds ratio. In a nested case-control study with cumulative sampling, the odds ratio estimates the relative risk of infection θ, provided the rare disease assumption is met. In a case-control study with density sampling, the parameter that can be estimated is the relative force of infection ϕ. If a source population member is enrolled in the study more than once, his data must be entered repeatedly.

Example 9.3 (continued) Assume that the study had a nested case-control design instead of a cohort design, and that at the end of the surveillance period a random sample from 400 controls was drawn, with $s_1 = 218$ and $s_0 = 182$. Then

$$OR = \frac{6 \times 182}{218 \times 33} = 0.152,$$

$$SE(\log OR) = \sqrt{1/6 + 1/218 + 1/33 + 1/182} = 0.455.$$

The lower and upper 95% confidence limits for the relative risk of infection θ are

$$LCL_{\theta} = \exp(\log 0.152 - 1.96(0.455)) = 0.062$$

and

$$UCL_{\theta} = \exp(\log 0.152 + 1.96(0.455) = 0.371.$$

Hence, the crude estimate of the oral cholera vaccine effectiveness ϑ is

$$VE = 1 - 0.152 = 0.848,$$

with as 95% confidence interval

$$(0.629, 0.938).$$

Here too, the estimator and its standard error can be transformed into a z-test:

$$z = \frac{\log(OR)}{SE(\log OR)},$$

the logit test for an odds ratio.

9.4.4 *Case-Cohort Data*

Recall that in a case-cohort study there can be overlap between the cases and the controls, and that this must be accounted for in the statistical analysis. A maximum likelihood approach for crude (and stratified) estimation in case-cohort designs was developed by Sato [70, 71].

The notation for case-cohort data is given in Table 9.3. Define $c_{1.} = c_{11} + c_{10}$, the total number of vaccinated cases, $c_{0.} = c_{01} + c_{00}$, the total number of unvaccinated cases, $c_{..} = c_{1.} + c_{0.}$, the total number of cases, and

$$s_1^* = \frac{c_{1.}c_{.0}}{c_{..}} + s_1$$

and

$$s_0^* = \frac{c_{0.}c_{.0}}{c_{..}} + s_0.$$

s_1^* and s_0^* are called pseudo-denominators. The maximum likelihood estimator of the relative risk of infection θ is

$$RR_{Sato} = \frac{c_{1.}/s_1^*}{c_{0.}/s_0^*}. \tag{9.2}$$

Table 9.3 Notation for crude case-cohort data

	Vaccinated	Unvaccinated	Total
Case but not control	c_{11}	c_{01}	$c_{.1}$
Case and control	c_{10}	c_{00}	$c_{.0}$
Control but not case	s_1	s_0	$s_.$

The formula for the standard error of $\log(RR_{Sato})$ is

$$SE(\log RR_{Sato}) = \sqrt{\frac{1}{c_{1.}} + \frac{1}{c_{0.}} + \left(1 - \frac{2c_{.0}}{c_{..}}\right)\left(\frac{1}{s_1^*} + \frac{1}{s_0^*}\right) - \frac{s_.^2 c_{1.} c_{0.} c_{.1} c_{.0}}{(c_{..} s_1^* s_0^*)^2}}.$$

Sato's chi-square test statistic to test the null hypothesis $H_0 : \theta = 1$ is

$$\chi^2 = \frac{N(c_{1.}s_0 - c_{0.}s_1)^2}{c_{..}(c_{1.} + s_1)(c_{0.}s_0)s_1},$$

where N is the total number of subjects enrolled in the study. If there is no overlap between the cases and the controls, Sato's method yields the same results as the Wald method for crude odds ratio analysis in Sect. 9.4.3.

9.5 Adjusted Data Analysis

As said, in the statistical analysis stage, confounding can be eliminated either by stratification, regression or propensity scoring.

9.5.1 Stratification to Adjust for Confounding

Stratification is a tried and trusted, robust method to adjust for confounding. In a stratified analysis, strata are created within which the confounder does not vary, or only minimally. Within each stratum, the confounder cannot confound because its level is the same for both vaccination groups. For each stratum separately the relative risk of infection for vaccinated subjects is estimated, and, if homogeneous, these stratum-specific estimates are then combined to obtain an overall, confounder-adjusted relative risk estimate.

For each of the statistical tests in Table 9.2, a stratified version exists. For case-control data, the appropriate stratified test is the Cochran–Mantel–Haenszel test for odds ratio analysis (also Mantel–Haenszel test). For cohort data, the appropriate stratified test is the Cochran–Mantel–Haenszel test for relative risk analysis. Both tests and corresponding estimators are available in PROC FREQ. A stratified Cox regression can be obtained by naming the covariate containing the strata in the CLASS statement. For the stratified version of Sato's test, see Refs. [70, 71].

Example 9.4 In seasonal influenza vaccine effectiveness studies, previous recent exposure, either through infection or vaccination, is a confounder. Exposure to influenza virus induces cellular and humoral immunity that protects against infection by the original infective strain, but also against antigenically similar strains. This type of protection is called *cross-protection*. Vaccination in a previous, recent season is a cause of vaccination in the current season, and the protection it induced in the previous season may lead to cross-protection in the current season. Persons who have been vaccinated previously are more likely to be vaccinated for the current season, and because antibodies induced by previous vaccinations may offer protection against current viruses by means of cross-protection, previous vaccination is a confounder. Jackson et al. report on influenza vaccine effectiveness in the United States during 2015–2016 season [72]. In their study, a test-negative design, they enrolled almost 7,000 patients six months of age or older who presented with acute respiratory illness at ambulatory care clinics in geographically diverse sites. In their data, previous vaccination is a confounder. Among those vaccinated in the previous season

$$\frac{1,634}{1,634 + 404} = 80.2\%$$

were vaccinated in the current season, compared to

$$\frac{740}{740 + 1,771} = 29.5\%$$

among those who were not vaccinated. In the group of previously vaccinated participants

$$\frac{48}{48 + 404} = 10.6\%$$

were influenza infected, compared to

$$\frac{272}{272 + 1,771} = 13.3\%$$

in the group of unvaccinated participants. Hence, vaccination in the previous season qualifies as a confounder, although not a very strong one. The relative risks of infection with the current virus were

$$\frac{162/48}{1,634/404} = 0.834$$

and

$$\frac{79/272}{740/1,771} = 0.695.$$

When vaccination in the previous season is ignored, the estimated relative risk of infection becomes

$$\frac{(162+79)/(48+272)}{(1,634+740)/(404+1,771)} = 0.690.$$

The SAS code for a stratified analysis according to the Cochran–Mantel–Haenszel method is

SAS Code 9.1 Stratified analysis of the data in Table 9.4

```
data;
   input Previous Current Case Count @@;
datalines;
1 1 1 162 1 1 0 1634 1 0 1  48 1 0 0  404
0 1 1  79 0 1 0  740 0 0 1 272 0 0 0 1771
;

proc freq order=data;
   table Previous*Current*Case /cmh;
   weight Count;
run;
```

SAS Output 9.1

```
          Estimates of the Common Relative Risk (Row1/Row2)

Type of Study  Method             Value      95% Confidence Limits
Case-Control   Mantel-Haenszel    0.7413      0.6025        0.9121
```

The Cochran–Mantel–Haenszel estimate of the common relative risk of infection for participants vaccinated for the current season is 0.7413. There is no statistical evidence that previous vaccination is an effect modifier (P-value for interaction 0.405.).

Stratified analysis works best if the number of confounders to adjust for is small. For each possible combination of confounder levels, a separate stratum must be created, with the risk of a large amount of sparsely populated strata with too little data to estimate the association between vaccination and prevention of infection with any reasonable degree of precision.

Table 9.4 TND effectiveness data against influenza A-H1N1 strain [72]

	Vaccinated in previous season			
	Yes		No	
	Current season		Current season	
	Vaccinated	Not vaccinated	Vaccinated	Not vaccinated
Cases	162	48	79	272
Controls	1,634	404	740	1,771

9.5.2 *Stratification to Adjust for Confounding by Calendar Time*

In Sect. 9.2.2, it was argued that when density sampling is applied, time-matching is required if the ratio of vaccinated members to unvaccinated members varies over the surveillance period. If this is ignored, confounding by calendar time may happen. The proper way to analyse time-matched case-control data is a stratified analysis, with each case and its time-matched controls constituting a stratum.

In case of a TND study, there is no time-matched sampling. Instead the controls are enrolled throughout the surveillance. But because the controls are not enrolled at random time points, here stratification by calendar week or calendar month is required as well. The reason is that most controls will be enrolled around the time of the peak occurrence of ARI cases. And just like in a case-control study with time-matched sampling, the ratio of these controls will estimate the ratio of the total person-times at the time of being enrolled rather than the ratio of the total person-times across the whole surveillance period (cf. Sect. 9.2.2).

9.5.3 *Regression to Adjust for Confounding*

A statistical method that can handle large numbers of confounders simultaneously is regression (Table 9.5). One can control for confounders like age, sex, ethnicity, health status, etc. in the same model, and by adding interaction terms, effect modification can be inspected. When the data are case-control data, logistic regression can be applied. The logistic model is

$$\text{logit } Pr(\text{case}|\text{x}_1, \ldots, \text{x}_\text{k}) = \beta_0 + \beta_1\text{x}_1 + \Sigma_{\text{i}=2}^{\text{k}} \beta_\text{i}\text{x}_\text{i},$$

where x_1 is a dichotomous variable coded as 1 for vaccinated cases and controls and as 0 for unvaccinated cases and controls, and the x_i are confounder regressions. The parameter β_0 is usually of no particular interest. The important parameter is β_1 because e^{β_1} is the vaccination odds ratio. Depending on the how the controls were sampled, the vaccination odds ratio is either the relative risk of infection for vaccinated subjects, in case of cumulative sampling

Table 9.5 Regression analysis by study design

Source population	Study design	Regression method	Effect measure
Cohort	Cohort	GLM analysis with link = log	θ
		Cox regression	ϕ
	Case-cohort	Logistic regression for case-cohort design	θ
	Case-control with cumulative sampling	Logistic regression	$\approx\theta$*
	With density sampling	Logistic regression	ϕ
Dynamic population	Case-control with density sampling	Logistic regression	ϕ

*If the rare disease assumption is met

$$e^{\beta_1} = \theta,$$

or the relative force of infection, in case of density sampling

$$e^{\beta_1} = \phi.$$

Recall the stratified case-control data in Example 9.4. The SAS code for the analysis of the data by means of logistic regression is

SAS Code 9.2 Logistic regression analysis of the data in Table 9.4

```
proc logistic descending;
   class Previous Current / param=ref ref=first;
   model Case(event="1")=Current Previous;
   freq Count;
run;
```

SAS Output 9.2

```
           Odds Ratio Estimates

                        Point         95% Wald
Effect               Estimate   Confidence Limits
Current 1 vs 0          0.745   0.607         0.915
```

Standard logistic regression can perform poorly when the number of parameters to be estimated is large, for example, in case of many strata. An alternative then is conditional logistic regression, an extension of logistic regression, and a very flexible procedure to take into account stratification (and matching). The approach conditions on the number of cases in each stratum, and by doing so eliminates the need to estimate the strata parameters. To obtain a conditional analysis of the data of Example 9.4, the MODEL statement in SAS Code 9.2 must be replaced with the following statement:

```
model case(event="1")=Current;
strata Previous;
```

9.5.4 Representation of Confounder Data

For categorical (dichotomous, nominal and ordinal) confounders, Miettinen advises what he calls 'liberal control', i.e. using indicator variates, because this approach maximizes the thoroughness of control [73].

Example 9.5 Assume that the health status has been scored using an ordinal scale with the following three categories: *good*, *fair*, *poor*. For this scale, the indicator variates are

x_1 = indicator of *fair*
x_2 = indicator of *poor*

In this model, *good* is the referent category.

Note that the quantitativeness of the information is ignored, because for the control of confounding this information is not relevant.

Incorrect modelling of a continuous confounder can result in residual confounding. If the association between the continuous confounder and risk of influenza infection is not linear, but, for example, U- or J-shaped, the assumption of a linear relation between the confounder and influenza infection can result in substantial residual confounding. Two sophisticated but challenging to interpret methods to model the relation between a continuous variable and an outcome are fractional polynomials and restricted cubic splines [74, 75]. However, it has been shown that adjustment for a continuous confounder by means of the much simpler method of stratification of the confounder in five strata and use of fractional polynomials or restricted cubic splines yield similar results [76].

9.6 Selecting Confounders in the Regression Model

A popular strategy for confounder selection is as follows [77]:

1. In the study protocol draw up a list of known confounders and a list of potential confounders. The list of known confounders can be based on the results of previous studies while the list of potential confounders can be based on expert knowledge.

2. All known confounders should be adjusted for, i.e. be included in the regression model.
3. Select possible confounders one by one in the model, based on the change-in-estimate criterion, i.e. at each step add that confounder that leads to the greatest change in the estimate of the relative risk (or the relative force of infection).
4. Stop adding variables to the model if the changes in the relative risk estimate become non-meaningful.

Popular choices for cut-offs for non-meaningful changes are 5 and 10%. Assume that the cut-off is set to 10%. Then, if after inclusion of the next confounder the relative risk estimate changes from, say, 0.34 to 0.32, the change ($(0.34-0.32)/0.34 = 5.9\%$) would be considered non-meaningful, and the adding of variables to the model would be terminated.

The above selection strategy may look straightforward and easy to apply, but, like all variable selection strategies, it can lead to difficult to solve problems. The first problem is that of sparse data bias, which has been called 'a problem hiding in plain sight'. This bias occurs when for some combinations of risk factors there are no or only a few infected cases, and it can occur even in quite large data sets. The symptom of sparse data bias is that the relative risk estimates get smaller and smaller, further away from the null (i.e. from 1.0), often with excessively wide confidence intervals. This should not be interpreted as evidence of confounding. A rule of the thumb is that confounder-adjusted estimates often only differ modestly from the unadjusted estimate. If sparse data bias is suspected, reducing the number of strata may be attempted. A consequence of sparse data bias may be that not all confounders can be controlled for, in which case a method called 'propensity scoring' could be attempted.

Another problem with the above selection strategy is that it like all stepwise analyses impacts the actual confidence level. This is true if the stepwise selection is taken into account. However, if only the final model is taken into account, the confidence level is not impacted.

9.7 Propensity Score Method

The propensity score method removes confounding caused by the observed covariates, by balancing baseline covariate values between vaccinated and unvaccinated subjects [78]. This is achieved by assigning each subject a so-called 'propensity score'. The propensity score is then the predicted probability of being vaccinated. Vaccine effectiveness estimates are obtained by adjusting for the propensity score as a linear or categorical variable or by matching subjects with similar propensity scores. While, in many cases, similar results will be obtained, there are important potential advantages to propensity scoring over conventional regression. For example, with propensity scoring one need not to be concerned with over-parametrization and can include non-linear terms and interactions. When attack rates are low but vaccination is common, propensity scoring may be a better option than logistic regression if many confounders must be adjusted for. Finally, propensity scoring tends to be the more robust method. For an example of a vaccine effectiveness with propensity scoring, see the publication by Simpson et al. [79].

Chapter 10
Meta-Analysis of Vaccine Effectiveness Studies

Abstract This chapter explores the meta-analysis of vaccine effectiveness studies. These meta-analyses tend to be non-comparative, although a simple statistical test to compare vaccine effectiveness estimates exists, as well as a powerful regression method known as meta-regression. It is demonstrated how these comparative analyses, as well as non-comparative analyses according to, for example, the DerSimonian–Laird method, can be conducted using simple SAS codes. It is shown that analytical confidence intervals for the difference of two vaccine effectiveness estimates do not exist, and that the problem of finding such a confidence interval comes down to finding a confidence interval for the difference of the medians of two lognormal distributions, and that such an interval can be obtained by means of parametric bootstrapping.

10.1 Introduction

When several independent studies on the effectiveness of a vaccine are available, there is usually a wish to systematically and quantitatively combine and summarize the results of the studies, i.e. for a meta-analysis. It is generally agreed that a meta-analysis provides a stronger conclusion than any of the individual studies. The benefit arises for two main reasons. First, combining results from similar studies will increase the precision of the vaccine effectiveness estimate. Second, meta-analyses allow to examine whether the vaccine effectiveness differs between populations or age groups, vaccine types, climates, etc.

A meta-analysis of vaccine effectiveness estimates involves summarizing log-transformed relative risk of infection estimates. Often these estimates will be odds ratios, when the individual studies are case-control or test-negative designs. When an individual study is a cohort design, the estimate will be a rate ratio or an incidence rate ratio. All three types of estimates can be combined in a meta-analysis, because all are estimates or approximate estimates of the relative risk of infection. To stress this, the estimates will be denoted by *RRE*, *relative risk estimate*. Thus, depending on the design of the individual study, *RRE* = *OR*, *RRE* = *RR* or *RRE* = *IRR*. The input data for the meta-analysis are the study-specific, confounder-adjusted relative

J. Nauta, *Statistics in Clinical and Observational Vaccine Studies*,
Springer Series in Pharmaceutical Statistics,
https://doi.org/10.1007/978-3-030-37693-2_10

risk estimates RRE_i and the standard errors $SE(\log RRE_i)$ of the log-transformed estimates. Following the example of Sutton et al., meta-analytical summary (pooled) estimates will be overlined: $\overline{RRE}$ [80].

The only meta-analysis readily available in SAS is the analysis according to the Cochran–Mantel–Haenszel method, which is available in PROC FREQ. Other analyses, including the one based on the popular DerSimonian–Laird method, must be programmed (but macros are available on Internet [81]). Fortunately, there is no need for extensive programming as it will be shown that all meta-analyses discussed in this chapter can done using PROC MIXED.

10.2 Non-comparative Analyses

There are two types of non-comparative meta-analyses: analyses based on a fixed-effect model and analyses based on a random effects model. In a fixed-effect model, it is assumed that the vaccine effectiveness is homogeneous and does not differ between studies. In contrast, in a random effects model, it is assumed that the vaccine effectiveness varies randomly between studies. In a fixed-effect meta-analysis, the only contributor to the variance of the summary estimate of the relative risk of infection is the within-study variance, the variance of the study-specific log-transformed RRE_i. In a random effects meta-analysis, there are two contributors to the variance of the summary estimate, the within-study variance and the between-study variance (see below).

The most applied random effects model is the DerSimonian–Laird model. Well-known fixed-effect methods are the Cochran–Mantel–Haenszel method and a modification of this method, the Peto method [82]. The methods have in common that summary estimate is a weighted average of the study-specific log-transformed RRE_i, with weights based on (the inverses of) the study-specific variances (Cochran–Mantel–Haenszel method, Peto method) or a combination of the study-specific variances and the between-study variance (DerSimonian–Laird method). The more precise the study-specific RRE_i, the more weight is given to the study, i.e. the larger the contribution of the study to the summary estimate. Because a random effects model incorporates between-study variance, an analysis based on a random effects model will generally yield a less precise summary estimate than an analysis based on a fixed-effect model.

Assuming that the vaccine effectiveness is a random effect comes down to assuming that the relative risk of infection varies randomly between studies, and that $\log \theta_i$, with θ_i the relative risk targeted in the ith study, is a random draw from a normal distribution with mean $\mu = \log \theta$ and variance σ^2, the between-study variance. When $\sigma^2 = 0$, the model becomes a fixed-effect model.

The probably most popular non-comparative random effects analysis is the one according to the DerSimonian and Laird method. In textbooks explaining the method, examples usually use raw data as input for the analysis, i.e. a set of study-specific four-fold tables. In the analysis, from these tables, the study-specific RRE_i and standard errors are calculated, and these are then summarized. This means that study-specific RRE_i can be used as input as well, and that there is no need for the raw data. This is

Table 10.1 Rotavirus vaccine effectiveness estimates [83]

Study	OR_i	95% CI	$SE(\log OR_i)$	Study	OR_i	95% CI	$SE(\log OR_i)$
1	0.20	(0.08, 0.52)	0.478	5	0.59	(0.19, 1.79)	0.572
2	0.17	(0.09, 0.32)	0.324	6	0.41	(0.23, 0.73)	0.295
3	0.44	(0.22, 0.88)	0.354	7	0.68	(0.44, 1.04)	0.219
4	0.23	(0.11, 0.49)	0.381	8	0.24	(0.14, 0.41)	0.274

important because it means that confounder-adjusted estimates can be used as input. Confounder-adjusted estimates can usually be taken from the publications selected for the meta-analysis. The steps are as follows:

1. Calculate from the study-specific vaccine effectiveness estimate VE_i the study-specific relative risk estimate as $RRE_i = 1 - VE_i$.
2. Convert the 95% confidence interval for VE_i into an interval for RRE_i.
3. Convert the confidence interval for RRE_i into an interval for $\log RRE_i$ by taking the logarithms of the limits in step 2.
4. Calculate $SE(\log RRE_i)$ by dividing the width of the confidence interval by $2 \times 1.96 = 3.92$.

Example 10.1 De Oliveira et al. report the results of a meta-analysis of rotavirus vaccine effectiveness against symptomatic infection in children under five in Latin American and Caribbean countries [83]. They report study-specific odds ratios rather than study-specific vaccine effectiveness estimates. In Table 10.1, their data are reproduced. Also shown are the $SE(\log OR_i)$. For example,

$$0.478 = \frac{\log 0.52 - \log 0.08}{3.92}.$$

A random effects analysis of the data can be performed using the following SAS code:

SAS Code 10.1 Random effects meta-analysis of log-transformed relative risks estimates

```
proc mixed method=ml;
   class Study;
   weight Wgt;
   model LogRRE=Int / cl noint ddf=1000;
   random Study;
   parms (1.0)(1.0) / hold=2;
run;
```

The variable `Wgt` must contain the inverses of the squared standard errors $SE(\log OR_i)$ in Table 10.1. The variable `Int` is an intercept variable which must be set to 1. P-values and confidence intervals provided by PROC MIXED are based on the t-distribution rather than on the standard normal distribution, a convention in meta-analyses. However, the degrees of freedom of the t-distribution can be specified with the DDF = option. If the degrees of freedom are set to a very large number, P-values and confidence intervals based on the standard normal distribution are obtained. For the intercept, the degrees of freedom cannot be specified. That is why the MODEL statement in SAS Code 10.1 contains the self-made intercept variable `Int` and the NOINT option (= no intercept). The model being fitted is

$$\log\theta = \beta_0 = \beta_{01} + \beta_{02}.$$

The first parameter, β_{01}, is the between-study effect and the second parameter, β_{02}, the within-study effect. The PARMS statement specifies initial values for the variances of the two parameters. HOLD = 2 specifies that the value for the variance of the second parameter must be kept fixed to the initial value, and that each study has its own within-study variance and is considered to be known. (The initial value for the second parameter is not arbitrary, it must be set to 1.0.) PROC MIXED returns an estimate for β_0 (the intercept), but not for β_{01} and β_{02}, because the separate parameters are not estimable. Because β_0 is the sum of a random and a fixed effect, β_0 is a random effect. Its estimate is

$$\overline{B}_0 = -1.099,$$

with (−1.457, −0.741) as 95% confidence interval. This corresponds to a summary relative risk estimate of

$$\overline{RRE} = \exp(-1.102) = 0.333,$$

with (0.233, 0.477) as 95% confidence interval, and to a summary vaccine effectiveness estimate of

$$\overline{VE} = 1 - 0.333 = 0.667,$$

with (0.523, 0.767) as 95% confidence interval for ϑ.

These results are similar to the results obtained with an analysis according to the DerSimonian–Laird method. The two analyses differ only in the estimate of the between-study variance: 0.177 (PROC MIXED) versus 0.202 (DerSimonian–Laird). If in SAS Code 10.1 in the PARMS statement the initial value for the first parameter is set to the DerSimonian–Laird estimate of the between-study variance, 0.202, and the HOLD = option is set to HOLD = 1, 2 (meaning that in the analysis both values must be kept fixed), the DerSimonian–Laird estimate is obtained (summary log-transformed relative risk estimate = −1.104).

When SAS Code 10.1 is run, the estimates returned are maximum likelihood estimates. One can also choose the restricted maximum likelihood estimates, by specifying METHOD = REML instead of METHOD = ML. Then the resulting estimate for the between-study variance is identical to the iterated DerSimonian–Laird estimate [84].

The SAS code for a fixed-effect analysis is as follows:

SAS Code 10.2 Fixed-effect meta-analysis of log-transformed relative risks estimates

```
proc mixed method=ml;
   weight Wgt;
   model LogRRE=Int / cl noint ddf=1000;
   parms (1.0) / hold=1;
run;
```

The summary log odds ratio is

$$\overline{B}_0 = -1.026,$$

with standard error 0.114. Because the between-study variance is assumed to be zero, this standard error is considerably smaller than the standard error obtained with the random effects analysis, 0.182.

10.3 Comparative Analyses

To compare two summary vaccine effectiveness estimates $\overline{VE}_1$ and $\overline{VE}_2$, the null hypothesis $H_0 : \vartheta_1 = \vartheta_2$ can be tested using the following test statistic:

$$z = \frac{\log \overline{RRE}_2 - \log \overline{RRE}_1}{\sqrt{SE(\log \overline{RRE}_1)^2 + SE(\log \overline{RRE}_2)^2}}.$$

The test statistic is approximately standard normally distributed. The null hypothesis being tested is $H_0' : \theta_2 = \theta_1$, but $H_0 : \vartheta_1 = \vartheta_2$ implies H_0' and vice versa, because $\vartheta_i = 1 - \theta_i$.

The analytical confidence interval that can be derived from z is an interval for the difference $\log \theta_2 - \log \theta_1$. The resulting interval can be converted into a confidence interval for the ratio θ_2/θ_1. But a confidence interval for θ_2/θ_1 cannot be converted into an interval for the difference $\vartheta_1 - \vartheta_2$ or for the ratio ϑ_1/ϑ_2. Thus for neither the vaccine effectiveness difference nor the vaccine effectiveness ratio, an analytical confidence interval exists.

Fortunately, a non-analytical confidence interval for the difference $\vartheta_1 - \vartheta_2$ can be obtained by applying a statistical method known as bootstrapping, a computer-based technique that relies on re-sampling. The best-known and most applied bootstrap method approaches from a participant study data set and the estimate of the parameter for which a confidence interval is required. A computerized algorithm draws, at random but with replacement, a very large number of so-called bootstrap samples from the original data set. Each bootstrap sample must be of the same size as the original data set and provides an bootstrap estimate of the parameter of interest. The entirety of bootstrap estimates forms an empirical sampling distribution for the original estimate. A confidence interval around the estimate can be obtained by the percentile method. In this method, the lower and upper limit of the 95% confidence interval are the 2.5th and the 97.5th percentile of the empirical sampling distribution. This bootstrap method is called non-parametric bootstrapping because no assumptions about the distribution of the data need to be made.

For a meta-analysis of a set of published studies, the original, participant data sets are typically not available, but study-specific estimates such as means and standard deviations are, and parametric bootstrapping can be applied. First, the study-specific estimates are summarized. The summary estimates are then treated as parameters of the distribution that gave rise to them. Next, a very large number of bootstrap samples are generated from these sampling distributions. These bootstrap samples form an empirical sampling distribution for the measure of interest. A confidence interval can be constructed by the percentile method. This bootstrap method is called parametric bootstrapping.

Both $\log \overline{RRE}_1$ and $\log \overline{RRE}_2$ are lognormally distributed, with means $\mu_1 = \log \theta_1$ and $\mu_2 = \log \theta_2$, or $e^{\mu_1} = \theta_1$ and $e^{\mu_2} = \theta_2$. The parameters e^{μ_1} and e^{μ_2} are the medians of the two lognormal distributions. Thus, to obtain a confidence interval for the difference $\theta_1 - \theta_2$, a parametric bootstrap interval for the difference of the medians of two lognormal distributions is needed. How such an interval can be obtained is explained in Appendix G.

Example 10.2 Darvishian et al. report the results of a meta-analysis on the effectiveness of seasonal influenza vaccines in elderly people [85]. For six different epidemic conditions, they present summary odds ratios with 95% confidence intervals, for two classes of vaccine matching state. (Influenza viruses are constantly changing due to antigenic drift. This can lead to antigenic differences between the circulating influenza viruses and the viruses included in the vaccine; the vaccine and circulating viruses may not be closely related. The degree of similarity or difference between the circulating viruses and the viruses in the vaccines is referred to as *vaccine match* and *vaccine mismatch*).

In the first four columns of Table 10.2, the results as given by Darvishian and colleagues are reproduced. Six epidemic conditions are distinguished, three with a low degree of virus circulation and three with a high circulation. For each condition, there are two summary relative risk estimates with corresponding confidence intervals, one for match and another for mismatch. The standard errors for the log-transformed summary relative risk estimated were derived from the confidence interval and are given in column 5. For example,

$$0.235 = \frac{\log 1.03 - \log 0.41}{3.92}.$$

In column 6, the summary vaccine estimates $VE_i = 1 - OR_i$ and the P-value are given. Not surprisingly, for all epidemic conditions, VE_{mismatch} is smaller than VE_{match}. The next three columns present the values for $VED_i = VE_{\text{mismatch}} - VE_{\text{match}}$, the z and the P-values, and the bootstrap 95% confidence intervals. For the epidemic condition widespread outbreaks, the comparison yields a statistically significant result (P-value $= 0.014$), while for the condition epidemic seasons the P-value is almost

Table 10.2 Pooled influenza vaccine effectiveness estimates by epidemic condition and match [85]

Epidemic condition	Match	$\overline{RRE}_i$	95% CI	$SE(\log \overline{RRE}_i)$	$\overline{VE}_i$	$\overline{VED}_i$	z/P-value	Bootstrap 95% CI
Non-epidemic seasons	Yes	0.65	(0.41, 1.03)	0.235	0.35	−0.22	−0.917/0.359	(−0.716, 0.251)
	No	0.87	(0.57, 1.32)	0.214	0.13			
Sporadic activity	Yes	0.69	(0.48, 0.99)	0.185	0.31	−0.23	−1.192/0.233	(−0.613, 0.148)
	No	0.92	(0.68, 1.25)	0.155	0.08			
Local activity	Yes	0.62	(0.28, 1.36)	0.403	0.38	−0.21	−0.517/0.605	(−1.135, 0.600)
	No	0.83	(0.38, 1.79)	0.395	0.17			
Epidemic seasons	Yes	0.48	(0.39, 0.59)	0.106	0.52	−0.16	−1.946/0.052	(−0.328, 0.001)
	No	0.64	(0.52, 0.78)	0.101	0.36			
Regional outbreaks	Yes	0.42	(0.30, 0.60)	0.177	0.58	−0.15	−1.255 0.216	(−0.397, 0.083)
	No	0.57	(0.41, 0.79)	0.167	0.43			
Widespread outbreaks	Yes	0.54	(0.46, 0.62)	0.070	0.46	−0.18	−2.458/0.014	(−0.334, −0.035)
	No	0.72	(0.60, 0.85)	0.085	0.28			

significant (0.052). Note that the bootstrap confidence intervals are consistent with the P-values: when the P-value is statistically significant, zero is not included in the interval, and vice versa.

For non-epidemic seasons, for example, the limits of the bootstrap confidence interval were estimated as follows: using SAS Code 10.3, 1,000,000 paired random draws (d_1, d_2) from the two normal distributions with

$$\mu_1 = \log 0.65 - \frac{1}{2}\sigma_1^2 \qquad \text{and} \qquad \sigma_1 = 0.235$$

and

$$\mu_2 = \log 0.87 - \frac{1}{2}\sigma_2^2 \qquad \text{and} \qquad \sigma_2 = 0.214$$

were generated. Next, for each random pair the bootstrap estimate $VED_{\text{bootstrap}}$ was calculated as

$$VED_{\text{bootstrap}} = \exp(d_1) - \exp(d_2).$$

The lower confidence limit for $\vartheta_1 - \vartheta_2$ was set to 2.5th percentile (−0.716) of the empirical distribution of the bootstrap estimates, the upper limit to the 97.5th percentile (0.251).

The bootstrap method cannot be used to obtain a confidence interval for the ratio ϑ_1/ϑ_2. The empirical distribution of the bootstrap estimates for ϑ_1/ϑ_2 is skewed, and the percentile method works well only if an estimator has a symmetric distribution. But see Chernick for adaptations of the method that try to overcome this [58].

SAS Code 10.3 Bootstrap confidence limits for a vaccine effectiveness difference

```
data;
   Sigma1=0.235; Mu1=log(0.65)-0.5*Sigma1**2;
   Sigma2=0.214; Mu2=log(0.87)-0.5*Sigma2**2;
   do Bootstrap_sample=1 to 1000000;
      D1=Mu1 + Sigma1*rannor(-1);
      D2=Mu2 + Sigma2*rannor(-1);
      VED=exp(D1)-EXP(D2);
      output;
   end;

proc univariate;
   var VED;
   output out=Out pctlpre=Percentile_ pctlpts=2.5, 97.5;

proc print;
run;
```

10.4 Meta-Regression

Meta-regression is weighted regression with studies as unit of observation rather than subjects. The dependent variable is the estimated vaccine effect in the study, and the weight is the inverse of the variance of the estimated effect. With this method, regression methods have become available to the meta-analyst, including the use of covariates. The covariates are also at study level. Both fixed-effect and random effects meta-regression are possible, as well as a combination of these two, mixed model meta-regression.

Early publications in which meta-regression was applied are a publication (1994) by Colditz et al. and a follow-up paper (1995) by Berkey et al. [86, 87]. Both authors use the same data set, the BCG data set. The Bacillus Calmette–Guérin (BCG) vaccine is a vaccine for the prevention of tuberculosis. The disease attacks the lungs and is caused by bacterium *Mycobacterium tuberculosis*. Tuberculosis, also known as 'consumption' or the 'white plague', was the cause of more deaths in industrialized countries than any other disease during the nineteenth and early twentieth century. By the late nineteenth century, 70–90% of the urban populations of Europe and North America were infected with the tuberculosis bacillus, and about 80% of those individuals who developed active tuberculosis died of it. Colditz and colleagues conducted a meta-analysis of 13 BCG vaccine efficacy studies. They developed a random mixed effects meta-regression model to enable exploration of sources of heterogeneity. Berkey and colleagues use the same data and absolute distance from the equator as surrogate for climate, to inspect the hypothesis that the BCG vaccine is more effective colder in climates. Latitude is a geographic coordinate that specifies the north–south position of a point on the earth's surface. It is an angle which ranges from 0° at the Equator to 90° at the poles. The BCG data are reproduced in Table 10.3. For a random effects meta-regression SAS Code 10.4 can be used.

Table 10.3 The BCG data set [84]

Study	RR_i	$SE(\log RR_i)$	Latitude	Study	RR_i	$SE(\log RR_i)$	Latitude
1	0.198	0.472	19°	8	0.625	0.238	27°
2	0.205	0.411	55°	9	0.712	0.111	18°
3	0.237	0.141	52°	10	0.804	0.226	13°
4	0.254	0.270	42°	11	0.983	0.267	33°
5	0.260	0.644	42°	12	1.012	0.063	13°
6	0.411	0.571	44°	13	1.562	0.730	33°
7	0.456	0.083	44°				

SAS Code 10.4 Mixed effects meta-regression analysis of the BCG data

```
proc mixed method=ml;
   class Study;
   weight Wgt;
   model LogRRE=Int Latitude / cl noint ddf=1000 ddf=1000;
   random Study;
   parms (1.0)(1.0) / hold=2;
run;
```

The model being fitted is

$$\log\theta = \beta_0 + \beta_1 \times \text{Latitude}.$$

The RANDOM statement specifies β_0 as a random effect. Its estimated value is

$$\overline{B}_0 = 0.281 \qquad (\pm\, 0.187).$$

The estimated value for β_1 is

$$\overline{B}_1 = -0.029 \qquad (\pm\, 0.0055).$$

The z-score, $-0.029/0.0055 = -5.27$, is highly significant. Thus, the higher the latitude, the smaller the relative risk of infection, and thus the higher the efficacy of the BCG vaccine.

Part IV
Correlates of Protection

Chapter 11
Immune Correlates of Protection

Abstract A correlate of protection is an immunological assay that predicts protection against infection. In clinical vaccine trials, correlates of protection are widely used as surrogate endpoints for vaccine efficacy. The function specifying the relationship between immunogenicity values and the probability of protection against infection is called the protection curve. It is demonstrated how the parameters of the protection curve can be estimated from both challenge study data and vaccine efficacy data. Also explained is how a threshold of protection can be estimated from the protection curve. The generalizable of estimated protection curves is discussed.

11.1 Introduction

An *immune correlate of protection* (correlate of protection for short) is an immune marker that predicts protection against infection or disease. (An immune marker is a measurable indicator of the immune state such as, for example, humoral or cellular immunological assays.) Correlates of protection are of great importance because they can be used as surrogate endpoints for vaccine efficacy in clinical vaccine trials. Immunological vaccine trials are much less costly and much less time-consuming than vaccine efficacy studies. Correlates of protection constitute the scientific basis for improving existing vaccines and introducing new ones. Under certain conditions, they can be used to predict vaccine efficacy in populations other than the one in which efficacy was demonstrated.

For many infectious diseases, it has been established that particular antibody assays are correlates of protection. Infectious diseases for which this has been demonstrated are, amongst others, meningococcal disease, influenza, hepatitis A and hepatitis B, tetanus, measles, mumps, rubella and polio.

In the scientific literature, the term correlate of protection is used in two different meanings. First, it is used in the meaning of an immune marker that predicts protection against infection or disease. Second, it is used in the meaning of a clear-cut value for an immunological assay, above which subjects are protected. In that case, often the protective level itself is called a correlate of protection. In this book, the term

J. Nauta, *Statistics in Clinical and Observational Vaccine Studies*,
Springer Series in Pharmaceutical Statistics,
https://doi.org/10.1007/978-3-030-37693-2_11

correlate of protection will be used in the first meaning only. For a protective assay level, the term *threshold of protection* will be used.

The correlation between an immune marker and protection from infection can be assessed in vaccine efficacy studies, challenge studies, household transmission studies and observational studies including cohort studies, natural history studies, maternal newborn studies, case-control studies and ecological studies [88].

11.2 Terminology for Correlates of Protection

Plotkin and Gilbert propose a nomenclature for correlates of protection [89]. They define a correlate of protection (CoP), as a marker of immune function that statistically correlates with protection after vaccination. A CoP may be either an mCoP, which is a mechanistic cause of protection, or an nCoP, which does not cause protection but nevertheless predicts protection through its (partial) correlation with another immune response(s) that mechanistically protects. In that case, the affection of the immune marker is a so-called collateral vaccine effect, completely unrelated to protection. To clarify their terminology, several examples are given. The protective effect of meningococcal vaccines is mediated by bactericidal antibodies, which are thus a Cop and an mCoP. The response to vaccination also induces antibodies that can be measured by an ELISA, but these antibodies do not cause protection. Thus the ELISA antibodies are a CoP and an nCoP. Another example concerns zoster vaccines. Zoster vaccines protect against herpes zoster (shingles), a painful skin rash with blisters. These vaccines were designed to elicit an immune response in older adults, whose immunity to varicella-zoster virus wane with advancing age. Both the antibody and cellular responses are correlated with protection and therefore, both are CoPs. But in view of the biology of the infection, the cellular response is an mCoP, whereas the antibody is an nCoP. Alternative terms for mCoPs and nCoPs are IM-1 and IM-2 markers (IM as abbreviation for immune marker) [88].

11.3 The Protection Curve

A *protection curve* is a mathematical function specifying the relationship between immune marker values and the probability of protection against infection, conditional on exposure to the pathogen. For convenience, it will be assumed that the immune marker values are antibody titres.

An obvious choice for the protection curve is the sigmoid logistic function:

$$f(t) = \frac{1}{1 + \exp(\alpha + \beta \log t)},$$

with $\alpha > 0$, $\beta < 0$ and t the antibody titre value. The standard logit function has two parameters, α and β, to be estimated from the data. The intercept α is the location parameter of the protection curve and reflects the protection not mediated

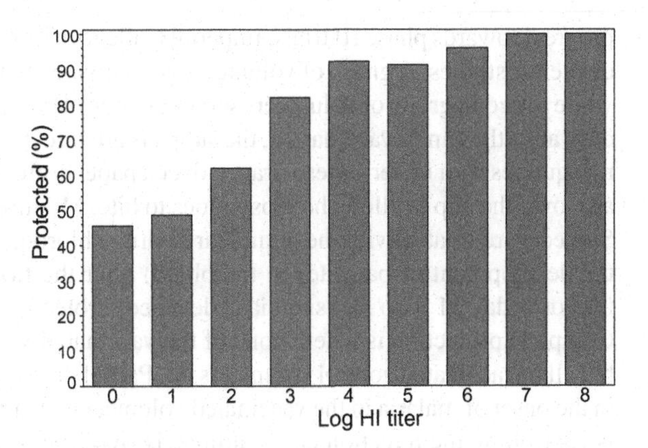

Fig. 11.1 Observed protection in an influenza vaccine challenge study

by antibodies but by other mechanisms. The slope β determines the steepness of the curve, the larger $|\beta|$, the steeper the curve. Steep protection curves are characteristic for all-or-nothing vaccines.

There are two criteria that must be satisfied for an antibody assay to be a correlate of protection. The first criterion is that $\beta \neq 0$. The second criterion, proposed by Qin et al., is the Prentice criterion [90]. In 1989, in what has become an influential article, Prentice formulated four criteria for the validation of surrogate endpoints [91]. The criteria ensure that rejection of the null hypothesis under the surrogate endpoint implies rejection of the null hypothesis under the true endpoint. The main criterion, often called the Prentice criterion, is that, independent of the treatment, there must be a single pathway from treatment to true endpoint through the surrogate endpoint. The Prentice criterion thus requires that the observed protective effect of the vaccine can be completely explained in a statistical model, by the immunological measurements only. This is not a given. If the control is a placebo, the antibody levels in the control group will be due to responding to natural infection, while the antibody levels in the investigational group will be due to responding to artificial infection. These antibody responses can be qualitatively different. This may also be the case if the control vaccine is a totally different type of vaccine (for example, intranasal) than the investigational vaccine. How it can be tested if the Prentice criterion is satisfied is explained in Sect. 11.4.2.

11.4 Estimating the Protection Curve

11.4.1 Estimating the Protection Curve from Challenge Data

A *challenge study* is a study in which vaccinated volunteers are challenged with a pathogen. Challenge studies are an important tool in clinical vaccine development, because they can furnish proof-of-concept for an experimental vaccine and accelerate progress towards phase III trials. Imperial College in London is famous for its malaria challenge studies. A group of volunteers is vaccinated with the experimental vaccine, while a second group of volunteers serves as control group. At a predefined number of days after the (final) vaccination, the subjects are infected with malaria. Five infected mosquitoes wait under a mesh draped over a paper coffee cup. The volunteer rests his arm over the cup to allow the mosquitoes to bite. Monitoring takes place twice daily. Subjects are treated with the antimalarial drug chloroquine (to prevent *parasitemia*, the development of parasites in the blood) after the first confirmed positive blood film or at day 21 if no parasitemia is detected. Protection can be complete or partial. Complete protection is where none of the vaccinated volunteers do develop malaria but all unvaccinated control volunteers do. Partial protection is where there is a delay in the onset of malaria in the vaccinated volunteers, meaning that the immune system is controlling the infection but is ultimately overwhelmed.

Challenge studies have a significant limitation, that it is often not possible to expose the volunteers to a wild-type strain, i.e. a strain found in nature, because that would be too dangerous. Instead, the volunteers are challenged with a laboratory-adapted strain, which is a strain weakened by passing. The limitation is the fidelity of a laboratory-adapted challenge model to natural infection.

Example 11.1 Hobson et al. studied the role of hemagglutination inhibition in protection against challenge infection with influenza viruses [2]. Four-hundred-and-sixty-two adult volunteers (industrial workers) were randomly assigned to be vaccinated with a live or a killed influenza vaccine or placebo, whilst others were left unvaccinated. Two to three weeks later they were challenged, by means of intranasal inoculation with a living, partially attenuated strains of an influenza B virus. Serum samples for anti-HA antibody determination by means of the HI test were drawn immediately before virus challenge. Nasal swabs were taken for virus isolation studies wherever possible 48 h after challenge. In total, 135 of the volunteers got infected and 327 remained infection-free. In Fig. 11.1, the observed proportions of protected subjects are shown.

SAS Code 11.1 Fitting a logistic protection curve to challenge data

```
proc nlmixed;
   parms Alpha=0.1 Beta=-0.1;
   Eta=exp(Alpha+Beta*Logtitre);
   P=1/(1+Eta);
   model Protected ~ binomial(1,P);
   predict Alpha+Beta*Logtitre out=Fitted;

data Pcurve; set Fitted;
   Pcurve=1/(1+exp(pred));
   Uclpcurve=1/(1+exp(lower));
   Lclpcurve=1/(1+exp(upper));
run;
```

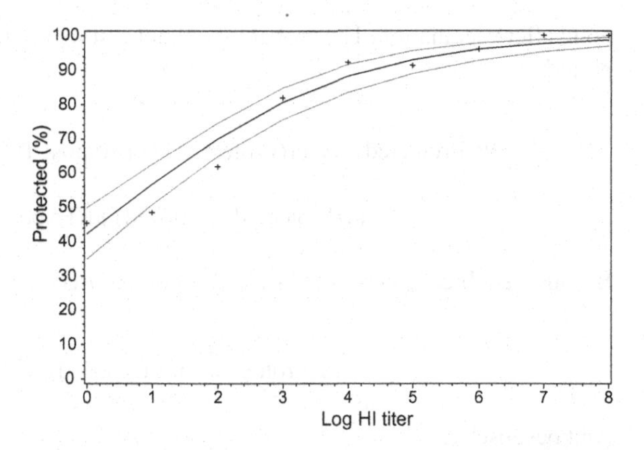

Fig. 11.2 Protection curve fitted to the challenge study data in Fig. 11.1

In SAS, there are several procedures that can be used to fit a logit protection curve to challenge data. The most flexible is PROC NLMIXED, see SAS Code 11.1. The variable `Logtitre` must contain the log-transformed antibody titres. The outcome variable `Protected` must be a binary variable, being set to 1 for subjects who were protected after challenge and to 0 for subjects who got infected. The variable `Pcurve` will contain the values for the fitted protection curve, and the variables `Lclpcurve` and `Uclpcurve` will contain the lower and upper limit of a two-sided 95% confidence interval for $f(t)$. Together, these confidence intervals constitute a 95% point-wise confidence band for the protection curve. The fitted protection curve and the confidence band are displayed in Fig. 11.2.

SAS Output 11.1

```
Parm   Estimate Std. Error   DF  t-Value  Pr > |t|  Alpha     Lower     Upper
Alpha    0.3130     0.1567  462     2.00    0.0463   0.05  0.005164    0.6209
Beta    -0.5794    0.06680  462    -8.67    <.0001   0.05   -0.7107   -0.4481
```

11.4.2 *Estimating a Protection Curve from Vaccine Efficacy Data*

A major difference between a challenge and a vaccine efficacy study is that in a challenge study all subjects are exposed to the pathogen, while in a vaccine efficacy study only a fraction of the subjects is exposed. This has to be allowed for in the model fitted to the data.

Consider a vaccine efficacy study with as source population a cohort and a fixed surveillance period, and with a particular antibody titre measured at a defined time-point after vaccination. The probability of not getting infected during the surveillance period is

$$\begin{aligned}\Pr(\text{Protected}) &= \Pr(\text{Protected} \mid \text{not Exposed}) \Pr(\text{not Exposed}) \\ &\quad + \\ &\quad \Pr(\text{Protected} \mid \text{Exposed}) \Pr(\text{Exposed}).\end{aligned}$$

Because a subject cannot get infected if not exposed,

$$\Pr(\text{Protected} \mid \text{not Exposed}) = 1.$$

And because

$$\Pr(\text{not Exposed}) = 1 - \Pr(\text{Exposed}),$$

the above equation can be rewritten as

$$\begin{aligned}\Pr(\text{Protected}) &= 1 - \Pr(\text{Exposed}) \\ &\quad + \\ &\quad \Pr(\text{Exposed}) \Pr(\text{Protected} \mid \text{Exposed}).\end{aligned}$$

It is the probability

$$\Pr(\text{Protected} \mid \text{Exposed})$$

that the interest is in, because this is the probability being modelled by the protection function. The expression above gives the model to be fitted.

Let P_E denote the probability that a subject is exposed to the pathogen, and $f(t)$ the protection curve, where t is the antibody titre value. Then

$$\Pr(\text{Protected} \mid t) = (1 - P_E) + P_E f(t).$$

If it is further assumed the $f(t)$ is the logistic function, then

$$\Pr(\text{Protected} \mid t) = (1 - P_E) + \frac{P_E}{1 + \exp(\alpha + \beta \log t)}.$$

This model is known as *Dunning's scaled logistic function* [92]. In the model exposure is modelled explicitly. The approach has been used to model the relationship between CMI responses to influenza vaccination in children and protection against

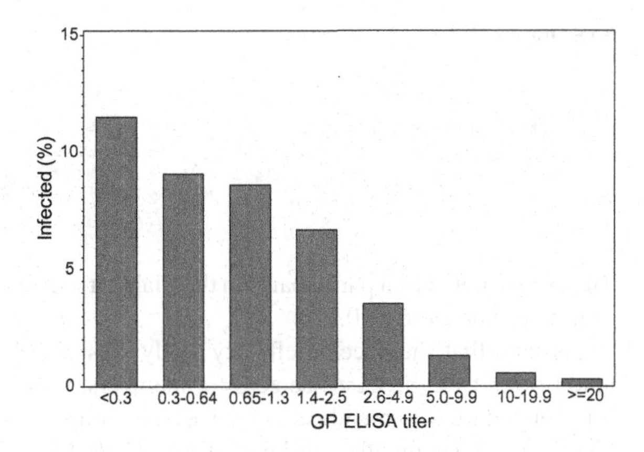

Fig. 11.3 Incidence of varicella in a vaccine efficacy study

culture-confirmed clinical infection with wild-type influenza virus [93]. Because the model separately parameterizes exposure, the protection curve can be estimated. This model too can be fitted with PROC NLMIXED.

Example 11.2 White et al. report the result of a vaccine efficacy study with a live attenuated varicella (chickenpox) vaccine conducted between 1987 and 1989 [94]. Four thousand forty-two healthy children and adolescents, aged 12 months to 17 years, were vaccinated with a single dose of the vaccine. During the first and second years of follow-up, 2.1 and 2.4% of the vaccines developed varicella. In Fig. 11.3, the incidence of varicella is shown by the 6-week post-vaccination glycoprotein-based (gp) ELISA assay titre. The following SAS code can be used to fit a protection curve to the field efficacy data in Fig. 11.3:

SAS Code 11.2 Estimating a logistic protection curve from vaccine efficacy data

```
proc nlmixed;
   parms Pe=0.1 Alpha=1 Beta=-1;
   Eta=exp(Alpha+Beta*Logtitre);
   P=(1-Pe)+Pe/(1+Eta);
   model protected ~ binomial(1,P);
   predict Alpha+Beta*Logtitre out=Fitted;

data Pcurve; set Fitted;
   Pcurve=1/(1+exp(pred));
   Uclpcurve=1/(1+exp(lower));
   Lclpcurve=1/(1+exp(upper));
run;
```

SAS Output 11.2

```
Parm Estimate Std. Error    DF t-Value  Pr > |t|  Alpha    Lower     Upper
Pe      0.1156    0.02561  3459    4.51    <.0001   0.05  0.06538   0.1658
Alpha  1.0182     0.7377   3459    1.38    0.1676   0.05  -0.4282   2.4647
Beta   -1.4812    0.2733   3459   -5.42    <.0001   0.05  -2.0171  -0.9453
```

The proportion of trial participants in the trial that was exposed to the varicella-zoster virus is estimated to be 0.116.

Assume that the vaccine efficacy study was placebo-controlled. Let the variable `Logtitre_active` be zero for members of the placebo group and the log-transformed titre for members of the vaccine group. Then, the Prentice criterion can be tested by changing the third line of SAS Code 11.2 to

```
Eta=exp(Alpha+Beta1*Logtitre + Beta2*Logtitre_active);
```

The Prentice criterion will be met if the estimated value of β_2 is close to zero.

11.4.3 Predicting Vaccine Efficacy

An estimated protection curve can be used to predict vaccine efficacy:

$$VE_{\text{predicted}} = 1 - \frac{\sum_{i=1}^{n_1} f(t_{1i})/n_1}{\sum_{i=1}^{n_0} f(t_{0i})/n_0},$$

where $t_{1\,i}$ is the antibody value of the ith subject in the vaccinated group and $t_{0\,i}$ the antibody value of the ith subject in the placebo group. This requires the assumption that the protection curve for placebo subjects is the same curve as that for vaccinated subjects. This need not to be so, however. In that case, the protection curve for placebo subjects should be estimated separately.

11.5 Threshold of Protection

Underlying the concept of a protection threshold is the idea that there is an assay value above which all subjects are protected from infection, and below which none are protected. This requires a very steep protection curve, a requirement that holds for all-or-nothing vaccines, but not for leaky vaccines. For leaky vaccines, an obvious definition of a protection threshold is the antibody value T_P for which the predicted probability of protection is 0.5:

$$f(t) = 0.5.$$

If the fitted protection curve is the logistic one, then it is easy to see that when the antibody titres are log-transformed, the estimate of T_P is the exponential of minus the ratio of the fitted intercept and slope:

$$\hat{T}_P = \mathrm{e}^{-\hat{\alpha}/\hat{\beta}}.$$

A confidence interval for T_P can be obtained using the PREDICT statement in PROC NLMIXED:

```
predict -Alpha/Beta;
```

The antilogs of the values for LOWER and UPPER then constitute a two-sided confidence interval for T_P.

Example 11.3 (continued) The estimate for the threshold of protection for varicella is a 6-week gp ELISA titre of

$$\exp(1.0182/1.4812) = 2.0,$$

with 95% confidence interval (0.9, 4.3).

Another method to estimate a threshold of protection is the *Chang–Kohberger method* [95]. To find a threshold of protection T_P, the following equation is solved:

$$\frac{\Pr(t < T_P \mid \text{Vaccinated})}{\Pr(t < T_P \mid \text{not Vaccinated})} = \frac{\Pr(\text{Infected} \mid \text{Vaccinated})}{\Pr(\text{Infected} \mid \text{not Vaccinated})}$$

Chang and Kohberger applied their method to aggregate efficacy data of trials with pneumococcal conjugate vaccine formulations and found a serotype 19F IgG antibody threshold of 0.4 μg/ml. The method requires (a) that the vaccine efficacy is known, and (b) that the antibody levels in the control group show substantial variability. In practice, only an estimate VE of the vaccine efficacy will be available, in which case the equation to solve becomes

$$\frac{\Pr(t < T_P|\text{Vaccinated})}{\Pr(t < T_P|\text{ not Vaccinated})} = 1 - VE.$$

If the control is a placebo and the post-vaccination antibody levels in this group are all undetectable or very low, then

$$\Pr(t < T_P|\text{not Vaccinated}) = 1.$$

Solving the equation above will not lead to a sensible result. But even if (b) is met, care should be taken. The Chang–Kohberger method is based on the assumption that the threshold T_P is the same for both the investigational and the control groups.

11.6 Correlates of Protection and Time

A protection curve models the probability of protection against infection or disease as a function of the immune marker levels. There are some subtleties to keep in mind, though, when estimating or interpreting a protection curve. If the immune marker is measured at a fixed time-point after vaccination, say, after three weeks, the marker is said to be a *fixed-time* correlate of protection. If marker levels do not decline over time, then the relationship between the levels and protection is time-independent, i.e. not dependent on the timing of the blood sampling. This is of importance for studying pathogenesis, which requires that the probability of protection is related to the antibody level at the time of exposure. If antibody levels, however, do decline over time, then the relationship will be time-dependent. If the antibody levels in the model were measured three weeks after the vaccination, the model may not be valid for antibody levels measured at later time-points. It is therefore of importance to always state clearly the timing of the blood sampling (for example, six weeks after vaccination). If the time-point at which the antibody titre is measured is the same for all trial participants, which is usually the case, then the (strength of the) relationship will depend on whether this point was before, at or after the time of the peak levels. The probability of protection will depend on the antibody level at the time of exposure. The lower the correlation between the measured antibody levels and the levels at the time of exposure, the weaker the relationship will be. Second, the relationship may be dependent upon the length of the surveillance period, when antibodies decline during the surveillance period.

If the marker is measured longitudinally, it is said to be a *time-dependent* correlate of protection. Fixed-time correlates of protection are used as predictors of protection against infection. Time-dependent correlates of protection are used to gain insight into mCoPs by relating current levels of the marker to the instantaneous risk of infection.

11.7 Generalizability

An important but difficult to answer question is the generalizability of estimated protection curves. For example, how generalizable are protection curves estimated from challenge data? This will depend, amongst others, on how similar the challenge strain is to the wild-type strain and on the volunteers, usually healthy young male and female students. Can a protection curve estimated from data collected in adults be assumed to be universal, for example, being applicable to an elderly population as well? Could the threshold be serotype dependent, or may it be different for some serotypes? Does it perhaps matter if antibodies are naturally or artificially acquired? The traditional threshold of protection for seasonal influenza vaccines, a haemagglutination inhibition titre of $\geq$40, is now being used for pandemic influenza vaccines as well, although there is little scientific justification for this.

Part V
Analysis of Vaccine Safety Data

Chapter 12
Vaccine Safety

Abstract To prove the safety of a vaccine is much more difficult than proving its efficacy. Of many vaccines, millions of doses are administered, which can bring very rare but serious adverse vaccine reactions to light. In this chapter, some major statistical aspects of vaccine safety are addressed. Vaccine safety surveillance is briefly discussed, and a recent vaccine safety controversy is recalled. The notorious problem of vaccine safety and multiplicity is explored at great length. Four different methods to handle this problem are presented, including the popular double false discovery method. The performance of the different methods is inspected with the help of simulation results. The second part of the chapter is dedicated to the analysis of local and systemic adverse reactions.

12.1 Ensuring Vaccine Safety

To prove the safety of a vaccine is much more challenging than proving its efficacy. Many vaccines are administered to several hundred million, often healthy people (e.g. childhood vaccines), in which case even extremely rare but serious adverse vaccine events can come to light, which may change the opinion of the medical community on the benefit/risk ratio. If a rare but serious condition occurs in, say, 0.1% of the population and a vaccine doubles the risk to 0.2%, then there will be additional 1,000 cases for every million persons vaccinated. A recent example of such an increased risk was a combination MMRV (measles, mumps, rubella, varicella) vaccine for children aged twelve months through twelve years, as alternative for two separate MMR and V vaccines. Post-licensure surveillance by the Vaccine Safety Datalink, a resource established by the United States Centers for Disease Control and Prevention (CDC) to investigate safety hypotheses using administrative databases of health maintenance organizations, detected a signal for increased febrile seizures in children between one and two years of age who had received the MMRV vaccine compared with those who had received the MMR vaccine. A febrile seizure, also known as fever fit or a fever convulsion, may happen with any condition that causes a sudden change in body temperature. These seizures can be caused by common childhood illnesses such as ear infection. During a febrile seizure, a child often has spasms and may

J. Nauta, *Statistics in Clinical and Observational Vaccine Studies*,
Springer Series in Pharmaceutical Statistics,
https://doi.org/10.1007/978-3-030-37693-2_12

lose consciousness. Vaccination may cause the body temperature to rise. It has been estimated that children who receive the combination MMRV vaccine are twice as likely to have a febrile seizure seven to ten days after the vaccination than children who get separate MMR and V vaccines. Because of this increased risk, MMRV is no longer advised over MMR + V separately.

To assess the causal link between a vaccine and a serious condition from observational data is extremely difficult. Hepatitis B vaccination has been linked to rheumatoid arthritis, lupus erythematosus, diabetes mellitus, acute leukaemia, chronic fatigue syndrome and hair loss, but none of this has been proven conclusively.

As another example, recall the emotionally charged thiomersal controversy. Thiomersal is an ethyl-mercury-containing preservative, which has been used to prevent bacterial contamination of vaccines since the 1930s. In 1999, the United States Food and Drug Administration (FDA) noticed that with the then vaccination program for children, infants, by the age of 6 months, could have received a total of 187.5 μg of mercury. This lead to concerns with the CDC and the American Academy of Pediatrics (AAP), the two organizations responsible for making childhood vaccine recommendations. Vaccine manufacturers were asked to remove thiomersal from their vaccines. This recommendation confused both parents and health care workers about the safety of vaccines. Studies were performed to investigate whether thiomersal in vaccines caused neuro-developmental or psychological problems. Evidence could not be found. Despite this, in 2000, the notion that thiomersal can cause autism emerged. This notion was disproved by several epidemiologic studies, examining the effect of reductions or removal of thiomersal from vaccines [96–99]. All these demonstrated that autism rates failed to decline despite removal of thiomersal, thus arguing against a causative role of thiomersal. The controversy has led to considerable medical and social damage.

Pre-licensure clinical vaccine trials typically focus on a special class of adverse events known as local and systemic reactions, on abnormal laboratory values and, depending on the vaccine, on abnormal vital signs values (body temperature and blood pressure), and on none-rare other adverse events. Phase I is mostly of an exploratory nature, to demonstrate initial safety, and often there is no statistical inference. Phase II is to quantify the occurrence of local and systemic reactions and laboratory abnormalities. Phase III is to evaluate less but non-rare common adverse events. Post-licensure the focus moves to rare but serious events by means of surveillance.

12.2 Vaccine Safety Surveillance

Vaccine manufacturers are required to report to the registration authorities all serious adverse events of which they become aware. Post-licensure (post-marketing) *vaccine safety surveillance* further relies on physicians and others to voluntarily submit reports of illness after vaccination. This is both the strength and the weakness of surveillance. The strength is that the system has proven to be able to detect very rare but serious risks of specific vaccinations. The weakness is that the reporting system has considerable limitations, including variability in the quality of the reports, biased reporting and underreporting, inadequate denominator data, absence of unvaccinated controls groups and the inability to determine whether a vaccine caused the adverse event in any individual report.

The problems vaccine surveillance is faced with are tremendous. Ellenberg gives a, what she calls, classic example of the problems of vaccine surveillance, that of *coincidental events* [100]. In the United States, sudden infant death syndrome (SIDS) during the first year of life occurs at a rate of about 1 in 1,300 infants. One can calculate, she writes, based on age-specific rates of SIDS and the current childhood vaccine schedule, that each year about 50–100 infants can be expected to die of SIDS within 2 days of being vaccinated.

Analysis of surveillance data is associated with statistical problems. Surveillance data contain strong biases. Reporting rates of specific adverse events cannot be calculated. Statistical significance tests and confidence intervals should be used with great reservation. If possible, safety signals should be confirmed in a randomized, controlled clinical trial. Menactra is a meningococcal conjugate vaccine. Meningococcal disease is a potentially fatal infection caused by bacteria (the meningococcal bacteria) that can infect the blood, the spinal cord and the brain. The vaccine contains four of the most common types of meningococcal bacteria and was licensed in the United States in 2005 for use in children and adults between the ages of 2 and 55 years old. In September 2005, the FDA and the Center for Biologics Evaluation and Research (CBER) revealed that officials had received five reports of Guillain Barré syndrome (GBS) connected to the Menactra vaccine, all in 17- and 18-year olds. GBS is a neurological disorder that can cause paralysis and permanent neurological damage. The majority of those affected recover, but recovery may take months and not infrequently may require hospitalization. GBS occurs when the immune system overreacts to foreign invaders. It can occur spontaneously and has been caused by infections, vaccinations, surgical procedures and traumatic injury. GBS was shown to have been a side effect of the swine influenza vaccine during the swine flu outbreak in 1976. To date, post-licensure surveillance did not reveal an association between vaccination with Menactra and GBS.

Post-licensure surveillance is sometimes criticized for underestimating benefits. Almost all children will have had a rotavirus infection by the age of 5. The virus is one of the most common causes of diarrhoea, which can be severe and dehydrating. In developing countries, rotavirus gastroenteritis is a major cause of childhood death. It has been estimated that the infection is responsible for approximately half a million deaths per year among children aged less than 5 years. Rotashield is a live attenuated rotavirus vaccine that was approved by the FDA in 1998. Little more than a year later, the manufacturer voluntarily withdrew it from the market. Shortly after approval, cases of intussusception were reported to the Vaccine Adverse Event Reporting System (VAERS), a surveillance system which collects information about possible side effects of licensed vaccines, a program of the FDA and CDC. Intussusception is a condition in which one bowel segment enfolds within another segment, causing obstruction. After licensure, VAERS recorded 76 cases, with 70% occurring after the first dose of the vaccine. The risk of intussusception has been estimated to be one case in every 5,000–9,500 vaccinated infants. Nonetheless, the vaccine could have prevented a considerable number of deaths in developing countries, where the benefit/risk rate would have been different. But because the vaccine was withdrawn from the United States market it could not be sold in developing countries.

A powerful statistical technique to investigate the association between unwanted events and transient exposure is the *self-controlled case series method*, or case series method for short. The method uses only data on cases, but it can provide estimates of the relative reporting rate of an adverse event. The method was developed to investigate a possible link between a MMR vaccine used in the United Kingdom and the occurrence of aseptic meningitis (an inflammation of the meninges caused by non-bacterial organisms). Strong evidence for a link between vaccination with the Urabe mumps strain and the disease was found, and several vaccines derived from this genotype mumps strain were withdrawn from the market. The case series analysis is based on conditional maximum likelihood estimation. For every case, a so-called case series likelihood is defined, and this likelihood is conditional on the case having occurred during the observation period. The observation period is split into successive intervals determined by changes in covariates and vaccine risk periods. For every period, a Poisson reporting rate is assumed. The case series likelihood is the case's contribution to the Poisson likelihood, conditioned on the case having occurred. The method controls for confounders that do not vary with time such as gender. In 1997, an intranasal influenza vaccine was granted approval for distribution and use in Switzerland. The nasal formulation consisted of an inactivated virosomal influenza vaccine, combined with a powerful mucosal adjuvant, heat-labile *Escherichia coli* enterotoxin. Shortly after the introduction, the vaccine was withdrawn from the market, because it was suspected that use of the vaccine increased the risk of Bell's palsy, a paralysis of the facial nerve leading to an inability to control facial muscles. Often the eye in the affected side cannot be closed and must be protected from drying up, to avoid permanent damage resulting in impaired vision. A report was published in the *New England Journal of Medicine* in which strong evidence for this increased risk was presented [101]. A strong relation in a case-control study was supported by a case series analysis that identified an increase in the rate of the condition, with a peak occurring between 31 and 60 days after intranasal vaccination followed by a return to the baseline level. For a comprehensive account of the self-controlled case series method, see the book by Farrington et al. [102].

12.3 Safety Data and the Problem of Multiplicity

The interpretation of safety data is complicated by the problem of multiplicity. The more safety variables are statistically analysed, the higher the false-positive rate (type I error rate) will be. Also, when the size of the safety database is large, clinically non-relevant differences will attain statistical significance. But any approach to control the false-positive rate will unavoidably decrease the false-negative rate (type II error rate). This could mean that some adverse vaccine effects may go undetected.

Several approaches to account for multiplicity in the analysis of safety data have been proposed. A first approach is to do nothing, to not correct. For many reviewers of safety data, false negatives (not rejected false safety null hypotheses) are of greater concern than false positives (rejected true safety null hypotheses). In that case, a conservative approach is not to adjust for multiplicity. This will increase the false-positive rate, but regulatory bodies are—or at least, should be—aware of this, and such safety signals (flaggings) are rarely a ground for a negative decision. Indeed, whereas regulatory agencies require multiplicity corrections for efficacy data, it is unlikely that they will accept such adjustment for safety data. Safety concerns may get special attention in post-licensure surveillance, or manufactures may be requested to perform a post-marketing safety study.

A second approach is to control the false-positive rate by a multiplicity adjustment that controls the family-wise error rate (FWER) in the strong sense, i.e. that controls the probability that at least one true safety null hypothesis is rejected. This can be achieved by applying, for example, the Bonferroni correction method, or the more powerful Holm method (also Bonferroni–Holm correction method). (For a detailed discussion on this correction methods, see the book by Dmitrienko, Tamhane and Bretz [103].) With the Bonferroni correction, if there are m null hypotheses, all hypotheses are tested at the significance level α/m. With the Holm correction, the m P-values are ordered such that

$$P_{(1)} \leq P_{(2)} \leq \cdots \leq P_{(m)}.$$

First, $H_{(1)}$ is tested at the level α/m. If $H_{(1)}$ is rejected, then $H_{(2)}$ is tested at the level $\alpha/(m-1)$. If $H_{(2)}$ is rejected, then $H_{(3)}$ is tested at the level $\alpha/(m-2)$, etc. If one of the hypotheses, say, $H_{(i)}$, cannot be rejected then no further null hypotheses are tested. This is why the Holm correction is called a step-wise correction method. But, as already noted, these multiplicity adjustments have the drawback that they increase the false-negative rate. For this reason, this approach is seldom applied in safety data analyses.

A third approach is the *false discovery rate* (FDR) *method*, introduced by Benjamini and Hochberg [104]. The FDR is defined as the expected proportion of rejected safety null hypotheses that are incorrectly rejected. Suppose that a safety analysis involves the statistical testing of 50 independent null hypotheses at the two-sided significance level 0.05, and that 40 of these null hypotheses are true and 10 false. Then the expected proportion of rejected true null hypotheses is $40 \times 0.05 = 2$, while the expected proportion of rejected false null hypotheses is $\sum_i Q_i$, with Q_i the probability that the ith false safety null hypothesis is rejected. If $Q_i = 0.90$ for all 10 false safety null hypotheses, then the expected number of true positives is $10 \times 0.90 = 9$. In that case, the FDR would be $2/11 = 0.18$. (The FDR will be approximately 0.18, because in the calculation the correlation between the numerator and denominator is ignored.) The false discovery rate approach aims to control the FDR at level α, by adjusting the significance level at which the safety null hypotheses are tested. When all safety null hypotheses are true, then the FDR procedure controls the family-wise error rate in the strong sense. But when some safety null hypotheses are false, then the statistical power of the FDR approach exceeds that of methods that control the FWER. Let

$$P_{(1)} \leq P_{(2)} \leq \cdots \leq P_{(m)}$$

be the ordered P-values for testing the safety null hypotheses

$$H_{(1)}, H_{(2)}, \ldots, H_{(m)}.$$

The FDR procedure rejects the j null hypotheses $H_{(1)}, H_{(2)}, \ldots, H_{(j)}$, where

$$j = \max\{i : P_{(i)} \leq (i/m)\alpha\}.$$

The adjusted P-values for the FDR procedure are

$$\text{adjusted } P_{(m)} = P_{(m)}$$

$$\text{adjusted } P_{(j)} = \min\{\text{adjusted } P_{(j+1)}, (m/j)P_{(j)}\}, \quad \text{for } j < m.$$

Consider the five ordered P-values in Table 12.1. All uncorrected P-values are <0.05, and thus all five safety null hypotheses $H_{(1)}, H_{(2)}, \ldots, H_{(5)}$ would be rejected. When the Bonferroni method is applied, all null hypotheses have to be tested at the $0.05/5 = 0.01$ significance level, and in that case only $H_{(1)}$ would be rejected. When the Holm method is applied, $H_{(1)}$ must be tested at the level $0.05/5 = 0.01$, $H_{(2)}$ at the level $0.05/4 = 0.0125$, and $H_{(3)}$ at the level $0.05/3 = 0.0167$. Because $P_{(3)} = 0.0212 > 0.0167$, $H_{(3)}$ cannot be rejected, and because the Holm method is a step-wise procedure, $H_{(4)}$ and $H_{(5)}$ can also not be rejected. With the Bonferroni method, only one null hypothesis would be rejected, while with the Holm method two null hypotheses would be rejected, which illustrates the difference in power between the two methods. On the last row of Table 12.1, the adjusted P-values for the FDR method are shown. All adjusted P-values are <0.05 and thus all five null hypotheses would be rejected.

To understand the differences between the uncorrected approach, the Holm method and the FDR approach, consider a study in which 45 independent safety null hypotheses are tested at the 0.05 significance level, and that 40 of these null hypotheses are true and 5 false, and that for each of these 5 null hypotheses the probability of a true-positive result is 0.90. Ten-thousand studies were simulated, and the results are shown in Table 12.2. If no corrections are made, the probability of at least one false positive (the probability of rejection at least one true safety null hypothesis) is as high as 0.873. The expected number of false positives is 2.0, while the expected number of true positives is 4.5. If the Holm method is applied, the probability that at least one true null hypothesis is rejected is 0.047. The expected number of false positives is 0.048, while the expected number of true positives is 2.3. Finally, if the

Table 12.1 Comparison of three approaches to control for multiplicity

Correction method	$P_{(1)}$	$P_{(2)}$	$P_{(3)}$	$P_{(4)}$	$P_{(5)}$
	0.0045*	0.0120	0.0212	0.0224	0.0493
Uncorrected	+	+	+	+	+
Bonferroni	+	−	−	−	−
Holm	+	+	−	−	−
FDR	+	+	+	+	+
	0.0225**	0.0280	0.0280	0.0280	0.0493

*unadjusted P-value; **adjusted P-value; + corresponding null hypothesis rejected; − null hypothesis not rejected

Table 12.2 Results of 10,000 simulated safety studies

Correction method	FWER[a]	Average number false positives	Average number true positives	FDR
No correction	0.875	2.0	4.5	0.281
Holm	0.047	0.05	2.3	0.017
FDR	0.169	0.20	3.0	0.045

[a] Average number of studies with at least one false positive (= a least one rejected true safety null hypothesis)

FDR method is applied, the probability of at least one false positive is 0.175. The expected number of false positives is 0.02, and the expected number of true positives is 3.0. The FDR is 0.045, a value which is indeed smaller than the significance level 0.05. The FDR method is thus a compromise between the uncorrected method—less false positives—and the Holm method—more true positives.

A fourth approach was proposed Mehrotra and Heyse, the *double false discovery rate* approach [105]. The double FDR approach is a two-step procedure for flagging adverse events. First, adverse events are grouped by body systems (e.g. the Body Systems of MedDRA, the standard medical terminology designed for the classification of medical information throughout the medical product regulatory cycle). Assume there are s body systems, and let P_{ik} be the P-value for testing H_{ik}, the kth safety null hypothesis of the ith body system. Then

$$P_i* = \min\{P_{i1}, P_{i2}, \ldots, P_{ik}\}$$

is the 'representative' P-value for the ith body system, i.e. the P-value for the strongest safety signal. The FDR procedure is applied to the P_i^* and, within the body systems, to the $P_{i1}, P_{i2}, \ldots, P_{ik}$. The double FDR procedure flags H_{ik} if

$$\text{adjusted } P_i^* < \alpha_1 \text{ and adjusted } P_{ik} \leq \alpha_2.$$

The authors advice to set α_1 to $\alpha/2$ and α_2 to α if the FDR is to be controlled at level α. The double FDR method substantially reduces the percentage of incorrectly flagged adverse events because it takes (some of) the dependency between events into account.

A fifth procedure that must be mentioned here was proposed by Berry and Berry [106]. Their approach is a Bayesian alternative to the double FDR approach, and since its publication in 2004 it has gained considerable popularity. Their model is a three-level hierarchical mixture model for simultaneously addressing many types of adverse events that are, like in the double FDR approach, grouped into body systems. The strength of the model is that it allows borrowing information both across- and within-body systems. Because of its complexity, the model is not discussed here. However, the publication may be of special interest to statisticians working in vaccine research because it presents a re-analysis of the vaccine safety data of Mehrotra and Heyse.

Events that regulators will rarely ignore are death serious autoimmune diseases, even if non-significant because of a multiplicity adjustment. They may also look at placebo-controlled trials and believe they have intuition about very rare events in the background of the population being studied and decide that 2 or 3 serious events of a particular type in a novel vaccine arm have to be investigated and not attributed to random chance.

12.4 Vaccine Reactogenicity

Vaccine *reactogenicity* refers to a set of common adverse events that are considered to be caused by or be attributable to the vaccination. They can be local or systemic. The reactions to be assessed depend on the type (class) of vaccine, its mechanism of action, route of administration, the targeted disease and the target population. Pre-licensing safety analyses typically focus on reactogenicity data.

Local reactions are reactions that occur at the site where the vaccine is administrated. In case of an injectable vaccine, these reactions are often called *injection site reactions*. They can be caused either by needle trauma or as an inflammatory reaction to the vaccine constituents. For example, local injection site pain is usually the consequence of some degree of tissue damage. Other examples of injection site reactions are impairment of arm movement, tenderness, erythema (redness), induration (swelling), itching and ecchymosis (blue spots). Common local reactions after nasal vaccination are nasal congestion and runny nose. A special class of local reactions are adjuvant-related local reactions, and the reader is referred to Chap. 18 of the book edited by Singh for a discussion on this topic [107].

Local reactions may be accompanied by *systemic reactions*, which are reactions that are the result of the immunological response to the vaccine. Typical examples of systemic reactions are headache, fever, malaise, fatigue, arthralgia (non-inflammatory joint pain), myalgia (muscle pain) and increased sweating. When the target population are toddlers, often graded systemic reactions are crying, irritability and decreased feeding.

Local and systemic reactions are usually collected with the help of a diary, which has to be filled in by the subject, or the parents in case of a childhood vaccine, for a period of 3 or 7 days after the vaccination. If reactions are predefined on the diary, they are called *solicited reactions* (in contrast to *unsolicited other adverse events*, which are collected on the adverse events pages of the case report form).

Severity of local and systemic reactions can be graded on a binary scale (*yes, no*), but more often an ordinal (ordered categorical) scale with the following 4 categories is used: *none, mild, moderate, severe*. A standard functional grading (categorization) is

mild: not interfering with normal daily activities,
moderate: interfering with normal daily activities and
severe: preventing one or more normal daily activities.

This grading makes the scale suited for rating reactions by subjects, which is an attractive property, but it may not be very sensitive to differences between vaccines. Also, it is not suited to grading a systemic reaction as fever. In an attempt to introduce uniform criteria for grading reactions, FDA/CBER in 2007 published grading scales for clinical and laboratory abnormalities for preventive vaccine clinical trials [108]. For erythema, the grades are based on the size of the greatest single diameter, while for induration (hardening of the skin) the grades are based on a functional assessment as well as an actual measurement. The FDA/CBER grades for the systemic reactions such as nausea and vomiting also take into account the number of episodes, while the grades for headache account for the use of nonnarcotic and narcotic pain relievers. The grades for diarrhoea are based on the frequency, shape and weight of the stools. (The FDA/CBER grading scales all contain a fourth grade: potentially life-threatening. Because this grade will rarely occur, it is usually omitted.)

The standard statistical analysis of reactogenicity data assesses the occurrence, the severity, the duration and, sometimes, the time after the vaccination of the local and systemic reactions.

The statistical analysis of local and systemic reactions usually starts with quantifying the reporting rate of the individual reactions by means of confidence intervals. Here, with reporting rate is meant the proportion of subjects reporting the reactions at least once during the 3 or 7 days follow-up period. The focus will usually be on the upper confidence limit for the rate, because it gives an upper bound for the rate with which the reaction is expected to occur among subjects receiving the vaccine. The bound is often translated into a less-than-1-in rate. If the upper confidence limit for the rate of a specific reaction is UCL, then the expected reporting rate of the reaction is <1 in $1/UCL$ vaccinated subjects, with $1/UCL$ often rounded down to the nearest multiplier of 100.

Example 12.1 Consider a vaccine safety database of 4,500 subjects who received the vaccine. Suppose that the systemic reaction sinusitis (inflammation of the paranasal sinuses) was reported by 3 vaccinees. The upper limit of the 95% Clopper–Pearson

Table 12.3 Reporting rates of 4 systemic reactions in a MMRV vaccines trial [105]

Systemic reaction	MMRV group ($n = 148$)	MMR + V group ($n = 132$)	Relative risk[a]
Malaise	27	20	1.16 (0.71, 2.08)
Constipation	2	0	1.80 P = 0.224
Diarrhoea	24	10	1.96 (1.08, 4.29)
Urticaria	0	2	0.00 P = 0.154

[a] with Jewell's correction

confidence interval for the reporting rate of sinusitis is 0.0019. Thus, the expected rate of sinusitis is <1 in 526 (i.e. <1 in 500) vaccinated subjects.

When reactogenicity experiences are to be compared between two vaccines—e.g. between an investigational vaccine and a control vaccine—it is done by computing confidence intervals for the relative risks of experiencing the reactions.

Example 12.2 In Table 12.3, the observed reporting rates of 4 selected systemic reactions reported by Mehrotra and Heyse (see Sect. 12.3) are given. Shown are the observed rates of the reactions malaise, constipation, diarrhoea and urticaria (hives) for the MMRV and the MMR + V vaccination groups. Also shown are the points estimates and two-sided 95% exact confidence intervals for the risk ratios, if existent, or the two-sided exact P-values.

Exact results are based on Barnard's test. The exact confidence intervals were found by trial-and-error using SAS Code I.2; the exact P-values were calculated using SAS Code 3.1.

Comparing reporting rates of local or systemic reactions between two vaccine groups is straightforward, but the disadvantage of comparing individual reactions between vaccine groups is that it does not allow accumulation of evidence. Such evidence could, for example, be that for all local or for all systemic reactions the rate ratio exceeded 1.0, but with none of the P-values being significant. Thus, although the data would strongly suggest that vaccine A is more reactogenic than vaccine B, there would be no statistical evidence to claim this. In that case, a simple but powerful approach is to analyse the intra-individual total numbers of local or systemic reactions. If on the diary there are, say, 6 solicited local reactions, then the intra-individual total number of local reactions can be 0, 1, 2, 3, 4, 5 or 6. Total numbers can be compared between two vaccine groups by means of Wilcoxon's rank-sum test.

Example 12.3 De Bruijn et al. compared the reactogenicity of a virosomal influenza vaccine to that of an MF59-adjuvanted vaccine in elderly [109]. The number of solicited local reactions on the diary was 8. Below the SAS analysis of the total numbers of local reactions is given. Note that the analysis is done using PROC FREQ (rather than PROC NPAR1WAY), using the CMH option with RANK scores. On average, the subjects vaccinated with the adjuvanted vaccine reported more local reactions than the subjects vaccinated with the virosomal vaccine. In the adjuvanted vaccine group, 43.2% of the subjects reported at least one local reaction, and 16.2% reported three or more reactions. In contrast, in the virosomal group only 21.2% of the subjects reported at least one local reaction while only 2 subjects reported 3 or more reactions. The statistic to compare the intra-individual total number is local reactions is Statistic 2, which is a chi-square statistic for a trend with 1 degree of freedom. For the example data, the two-sided P-value is <0.001, which allowed the conclusion that in elderly the adjuvanted influenza vaccine is more reactogenic than the virosomal vaccine with respect to local reactions.

SAS Code 12.1 Comparing intra-individual numbers of local reactions

```
proc freq;
   table Vaccine*Nreactions / cmh scores=rank;
run;
```

SAS Output 12.1

```
                        Nreactions
Vaccine         0      1      2      3      4      5      6      7 Total
-----------------------------------------------------------------------
Adjuvanted     70     26     13      8      3      5      1      4   130
             53.8   20.0   10.0    6.2    2.3    3.8    0.8    3.1
Virosomal     100     18      7      1      0      0      0      1   127
             78.7   14.2    5.5    0.8    0.0    0.0    0.0    0.8
-----------------------------------------------------------------------
Total         170     44     20      9      3      5      1      5   257

   Cochran--Mantel--Haenszel Statistics (Based on Rank Scores)
Statistic    Alternative Hypothesis    DF        Value        Prob
    2        Row Mean Scores Differ     1      21.1839      <.0001
```

As said, local and systemic reactions are usually scored for a period of 3 or 7 days after the vaccination. In that case, the intensity of the reaction is usually taken to be the maximum score during the follow-up period.

Local and systemic reaction ordinal scores can be compared between two vaccine groups by means of Wilcoxon's rank-sum test with modified ridit scores as ranks [110]. The category *none* is given grade 0, the category *mild* grade 1, the category *moderate* grade 2 and the category *severe* grade 3. The null hypothesis tested is that mean scores do not differ between the groups. This comparison can be done with PROC FREQ of SAS, using the CMH option with MODRIDIT scores.

Example 12.4 According to the results of a randomized study published in the *British Medical Journal*, longer needles for infant immunizations may cause fewer local reactions [111]. Compared with short narrow needles, use of long wide needles was associated with significantly decreased local reactions to diphtheria, tetanus, whole-cell pertussis and *H. influenzae* type b vaccinations. Significantly fewer infants vaccinated with the long needle had severe local reactions. Consider a (hypothetical) randomized trial comparing administration of a diphtheria vaccine using either a long (25 mm) needle or short (16 mm) needle. Suppose that local reactions were graded by parents trained how to do so, and that for the local reaction tenderness the results were as follows: infants vaccinated with the long needle: *none*: 30, *mild*: 20, *moderate*: 12, *severe* tenderness: 5; infants vaccinated with the short needle: *none*: 19, *mild*: 15, *moderate*: 19, *severe*: 10. To analyse these reaction scores, the SAS code 12.2 can be used. Here also, the statistic to use is Statistic 2. For the example data, the two-sided P-value is 0.0183. In the database, the scores must be ordered, e.g. 0, 1, 2, 3 or A, B, C, D but not *none, mild, moderate, severe*, because in that case PROC FREQ ranks the scores alphabetically, in which case an incorrect P-value is returned: 0.7272.

SAS Code 12.2 Comparing ordinal reaction scores

```
proc freq;
   table Vaccine*Score / cmh scores=modridit;
run;
```

Appendix A
SAS and Floating-Point Format for Calculated Variables

When using SAS, floating-point format for calculated variables should be avoided, especially when values are to be compared with a constant. As shown below, it may lead to errors. The solution to this problem is rounding. When calculating a value, use the function ROUND at the final step and round to, say, three decimals more than needed for the comparison. (But do not round to soon.) As an example, consider a trial in which every serum sample is titrated twice, with the titre assigned to the sample the geometric mean of the two assay values. Let the endpoint be whether or not the subject is seroprotected, for example, whether or not the assigned titre is greater than or equal to 40. With the floating-point format, errors will occur.

SAS Code A.1

```
data;
   input Subject Assay1 Assay2;
   Titre=exp((log(Assay1)+log(Assay2))/2); /* geometric mean */
   Titre_r=round(Titre,.001);              /* rounded titre */
   Sp=(Titre ge 40);                       /* seroprotected yes/no */
   Sp_r=(Titre_r ge 40);                   /* seroprotected yes/no derived */
datalines;                                 /* from rounded titre */
1 40  40
2 20  80
3 10 160
4  5 320
;

print; run;
```


J. Nauta, *Statistics in Clinical and Observational Vaccine Studies*,
Springer Series in Pharmaceutical Statistics,
https://doi.org/10.1007/978-3-030-37693-2

SAS Output A.1

Subject	Assay1	Assay2	Titre	Titre_r	Sp	Sp_r
1	40	40	40	40	1	1
2	20	80	40	40	0	1
3	10	160	40	40	1	1
4	5	320	40	40	1	1

All assigned titres should be 40, and for all four subjects both the non-rounded and the rounded calculated titres are printed as 40. But, when the calculated titre is not rounded, according to SAS, subject 2 is not seroprotected. This is due to the use of the floating-point format. When the values are rounded, this error does not occur.

Appendix B
Closed-Form Solutions for the Constrained ML Estimators $\tilde{R}_0$ and $\tilde{R}_1$

The standard errors (3.8) and (3.12) involve constrained maximum likelihood estimators $\tilde{R}_0$ and $\tilde{R}_1$ of the rates π_0 and π_1 [7]. For the standard error (3.8), the constraint is

$$\tilde{R}_1 = \tilde{R}_0 + \Delta_0.$$

Let c_0 and c_1 be the observed numbers of events, n_0 and n_1 the group sizes, $c = c_0 + c_1$ and $n = n_0 + n_1$. Define:

$$\begin{aligned}
L_0 &= c_0\Delta_0(1 - \Delta_0)\\
L_1 &= (n_0\Delta_0 - n - 2c_0)\Delta_0 + c\\
L_2 &= (n_1 + 2n_0)\Delta_0 - n - c\\
L_3 &= 3n.
\end{aligned}$$

The closed-form solution for $\tilde{R}_0$ is

$$\tilde{R}_0 = 2p\cos(a) - L_2/L_3,$$

where

$$\begin{aligned}
a &= (1/3)[3.1416 + \cos^{-1}(r)]\\
r &= q/p^3\\
q &= (L_2/L_3)^3 - L_1L_2/(6n^2) + L_0/(2n)\\
p &= \text{sign}(q)\sqrt{(L_2/L_3)^2 - L_1/L_3}.
\end{aligned}$$

When $q = 0$ then $p = 0$, and vice versa. In that case r must be set to 1.

J. Nauta, *Statistics in Clinical and Observational Vaccine Studies*,
Springer Series in Pharmaceutical Statistics,
https://doi.org/10.1007/978-3-030-37693-2

For the standard error (3.12), the constraint is

$$\tilde{R}_1 = \theta_0 \tilde{R}_0.$$

The closed-form solution for $\tilde{R}_0$ is

$$\tilde{R}_0 = \frac{-B - \sqrt{B^2 - 4AC}}{2A},$$

where

$$\begin{aligned} A &= n\theta_0 \\ B &= -(n_1\theta_0 + c_1 + n_0 + c_0\theta) \\ C &= c. \end{aligned}$$

Appendix C
Proof of Inequality (3.17)

Consider a trial with $k > 1$ primary endpoints, and with the objective to demonstrate that an experimental vaccine is superior (or non-inferior) to a control vaccine for *all* primary endpoints. Let E_i be the event that the trial yields a significant result for the ith endpoint, $P_i = \Pr(E_i)$ the statistical power of the trial for the ith endpoint, and $P = \Pr(E_1 \cap ... \cap E_k)$ the overall statistical power, i.e. the probability that the trial yields a significant result for all k endpoints. Then, the following inequality holds:

$$P \geq \sum_{i=1}^{k} P_i - (k-1).$$

Proof The inequality can be proven by mathematical induction. According to the addition rule for probabilities,

$$\begin{aligned} P &= \Pr(E_1 \cap E_2) \\ &= \Pr(E_1) + \Pr(E_2) - \Pr(E_1 \cup E_2) \\ &\geq \Pr(E_1) + \Pr(E_2) - 1 \\ &= P_1 + P_2 - (2-1). \end{aligned}$$

Thus, the inequality holds for $k = 2$. Assume that it has been shown that the inequality holds for $k = 2, \ldots, j$, with $j \geq 2$. Then, for $k = (j+1)$ it follows that

$$\begin{aligned} P &= \Pr(E_1 \cap \cdots \cap E_j \cap E_{j+1}) \\ &= \Pr(E_1 \cap \cdots \cap E_j) + \Pr(E_{j+1}) - \Pr((E_1 \cap \cdots \cap E_j) \cup E_{j+1}) \\ &\geq \Pr(E_1 \cap \cdots \cap E_j) + \Pr(E_{j+1}) - 1 \\ &\geq P_1 + \cdots + P_j - (j-1) + \Pr(E_{j+1}) - 1 \\ &= P_1 + \cdots + P_{j+1} - j \\ &= P_1 + \cdots + P_{j+1} - [(j+1) - 1]. \end{aligned}$$

□

J. Nauta, *Statistics in Clinical and Observational Vaccine Studies*,
Springer Series in Pharmaceutical Statistics,
https://doi.org/10.1007/978-3-030-37693-2

Appendix D
A Generalized Worst-Case Sensitivity Analysis for a Single Seroresponse Rate for Which the Confidence Interval Must Fall Above a Pre-specified Bound

D.1 Introduction

In 2007, FDA/CBER published two guidance documents for the licensure of influenza vaccines, one for seasonal inactivated vaccines and another for pandemic vaccines [51, 112]. Both documents give the same criteria for influenza vaccine immunogenicity. For an adult population, the lower limit of the two-sided 95% confidence interval for the seroprotection rate must meet or exceed 0.7, and the lower limit of the confidence interval for the seroconversion rate must meet or exceed 0.4. For an elderly population, the respective bounds are 0.6 and 0.3. Seroprotection and seroconversion are both binary outcomes. Seroprotection is defined as achieving an antibody level above a given threshold value. The standard definition of seroconversion is going from a pre-vaccination state of no detectable antibodies (seronegative) to a post-vaccination state of detectable antibodies (seropositive). An alternative definition of seroconversion is a significant post-vaccination increase in antibody level.

In case of a statistical analysis aimed at demonstrating that the confidence interval of a rate is above a pre-specified bound, the most applied method to handle missing data is, probably, the complete-case analysis. This analysis requires the assumption that the probability that an outcome is missing is independent of the outcome, i.e. the assumption that the probability that the outcome is missing does not depend on whether the outcome is positive (success, e.g. subject seroconverted) or negative (failure). A sensitivity analysis is an analysis that investigates the influence of deviations from the assumptions underlying the main analysis. For binary outcomes, a simple sensitivity analysis in case of missing data is to treat all subjects with a missing outcome as failures, and then to check if this analysis supports the conclusion from the complete-case analysis. The worst-case sensitivity analysis is based on an extreme assumption that only failures will be missing, and the more missing data there are, the more extreme the assumption is.

Here, a generalized worst-case sensitivity analysis for a single rate for which the confidence interval must fall above a pre-specified bound is proposed, based

J. Nauta, *Statistics in Clinical and Observational Vaccine Studies*, Springer Series in Pharmaceutical Statistics, https://doi.org/10.1007/978-3-030-37693-2

on the maximum likelihood (ML) method. The analysis checks for a continuum of assumptions, from the assumption underlying the complete-case analysis to the one underlying the worst-case analysis, if the bound lies within or outside the confidence interval.

D.2 Motivating Example

As a motivating example, consider a study in which 100 adult subjects are vaccinated with a pandemic A-H1N1 influenza vaccine. Suppose that three weeks after the vaccination 49 subjects have seroconverted and 41 not, and that for 10 subjects the outcome is missing. In the complete-case analysis, the FDA/CBER criteria for seroconversion are met because the lower limit of the 95% Clopper–Pearson confidence interval for the probability of seroconversion is 0.436, which exceeds the bound set by the Agency, $B = 0.4$. A sensitivity analysis in which all subjects with a missing outcome are assumed to have not seroconverted, however, does not support the conclusion of the complete-case analysis because in that analysis the lower limit of the Clopper–Pearson confidence interval is 0.389.

D.3 Complete-Case and Worst-Case ML Analyses

As said, the generalized worst-case sensitivity analysis proposed here is based on the ML method. Therefore, as an introduction, first the ML analyses of the complete-case and the worst-case data are described.

Let π denote the probability of a positive outcome, and ψ_s and ψ_{ns} the probabilities that a positive or a negative outcome is missing. In Table D.1, a probability model for the data including missing values is given. Note that it is assumed that the missing data mechanism depends on the outcome but not on any observed or non-observed covariate. The log-likelihood function for the data set is

$$\begin{aligned} LL(\pi, \psi_s, \psi_{ns}) = \; & s \log[(1 - \psi_s)\pi] \; + \; (m - s) \log[(1 - \psi_{ns})(1 - \pi)] \\ & + (n - m) \log[\psi_s \pi + \psi_{ns}(1 - \pi)], \end{aligned} \tag{D.1}$$

Table D.1 Probability model for binary observations with missing data

Event	Probability
Observed positive outcome	$(1 - \psi_s)\pi$
Observed negative outcome	$(1 - \psi_{ns})(1 - \pi)$
Missing observation	$\psi_s \pi + \psi_{ns}(1 - \pi)$

where s is the observed number of positive outcomes, m the total number of subjects with a non-missing outcome and n the total number of subjects.

The complete-case analysis requires the assumption that $\psi_s = \psi_{ns} = \psi$. In that case, the likelihood function (D.1) becomes

$$LL(\pi, \psi) = [s \log \pi + (m - s) \log(1 - \pi)] + [(n - m) \log \psi + m \log(1 - \psi)].$$

The first component of this log-likelihood function depends only on π while the second component depends only on ψ. Thus, both components can be maximized independently of each other. If the parameter of interest is π, then the second component is a constant and can be dropped from the log-likelihood function, which then simplifies to the log-likelihood function for complete-case data:

$$LL_{CC}(\pi) = s \log \pi + (m - s) \log(1 - \pi).$$

The null hypothesis H_0: $\pi = \pi_0$ can be tested using the likelihood ratio statistic:

$$LRS_{CC}(\pi_0) = 2[LL_{CC}(\hat{\pi}) - LL_{CC}(\pi_0)],$$

where $\hat{\pi}$ is the ML estimate of π. Under the null hypothesis, for large sample sizes, this statistic has a chi-square distribution with one degree of freedom. The likelihood ratio statistic can be used to derive a confidence interval for π. Any value π_0 for which $LRS_{CC}(\pi_0)$ is less than $\chi^2_{1-\alpha}$ is in the $100(1 - \alpha)\%$ likelihood-based confidence interval, and vice versa.

For the complete-case data, $\hat{\pi}$ is 49/90 = 0.544, with

$$LL_{CC}(\hat{\pi}) = -62.027.$$

For the FDA/CBER bound for seroconversion, the log-likelihood equals

$$LL_{CC}(0.4) = -65.842.$$

Thus,

$$LRS_{CC}(0.4) = 2(-62.027 + 65.842) = 7.630,$$

a value which exceeds $\chi^2_{0.95} = 3.841$. This implies that B is not in the 95% confidence interval. The lower confidence limit has to be found by iteration.

$$LRS_{CC}(0.442) = 3.796 < 3.841,$$

and

$$LRS_{CC}(0.441) = 3.971 > 3.841.$$

Thus, for the complete-case data, the lower likelihood-based confidence limit for the probability of seroconversion is 0.442, which is in good agreement with the Clopper–Pearson limit.

The log-likelihood function for the worst-case data, i.e. for the data set with the missing values replaced by zeros (failures), is

$$LL_{WC}(\pi) = s \log \pi + (n - s) \log(1 - \pi). \tag{D.2}$$

This is the same log-likelihood function as for the complete-case data, except that in the second term the multiplier $(m - s)$ is now $(n - s)$.
For the worst-case data, $\hat{\pi}$ is $49/100 = 0.490$, with

$$LL_{WC}(\hat{\pi}) = -69.295,$$

$$LL_{WC}(0.4) = -70.950,$$

and

$$LRS_{WC}(0.4) = 3.311 < 3.841.$$

Again, the likelihood analysis is in agreement with the Clopper–Pearson analysis that for the worst-case data the FDA/CBER bound B is not below but in the 95% confidence interval.

D.4 Maximum Likelihood Analysis with Missing Data

With the following re-parameterization: $\tau = \psi_{ns}/\psi_s$ and $\psi = \psi_{ns}$, the log-likelihood function (D.1) becomes

$$\begin{aligned} LL(\pi, \psi, \tau) = {} & s \log[(1 - \psi/\tau)\pi] + (m - s) \log[(1 - \psi)(1 - \pi)] \\ & + (n - m) \log[(\psi/\tau)\pi + \psi(1 - \pi)]. \end{aligned}$$

Let $\tilde{\pi}_\tau$ and $\tilde{\psi}_\tau$ denote the constrained ML estimates of π and ψ for τ fixed. The conditional null hypothesis H_0: $(\pi = \pi_0|\tau)$ can be tested using the conditional likelihood ratio statistic:

$$\text{CLRS}(\pi_0|\tau) = 2[LL(\tilde{\pi}_\tau, \tilde{\psi}_\tau, \tau) - LL(\pi_0, \tilde{\psi}_{0\tau}, \tau)], \tag{D.3}$$

where $\tilde{\psi}_{0\tau}$ is the constrained ML estimate of ψ under the conditional null hypothesis. The statistic $\text{CLRS}(\pi_0|\tau)$ can be considerably simplified, because of the interesting property that $\tilde{\psi}_{0\tau} = \tilde{\psi}_\tau$. A proof of this property is given below (see Sect. D.7). Because of this property, an alternative formula for the statistic is

$$\text{CLRS}(\pi_0|\tau) = 2[LL'(\tilde{\pi}_\tau, \tau) - LL'(\pi_0, \tau)], \tag{D.4}$$

where

$$LL'(\pi, \tau) = [s \log \pi + (m - s) \log(1 - \pi) + (n - m) \log(\pi/\tau + 1 - \pi)]. \quad \text{(D.5)}$$

Form. (D.4) is much easier to evaluate than Form. (D.3) because it does not involve $\tilde{\psi}_\tau$. Furthermore, for $\tilde{\pi}_\tau$ a closed-form solution exists (see Sect. D.7). Thus, to evaluate CLRS$(\pi_0|\tau)$, no (iterative) maximization is required.

Under the conditional null hypothesis, CLRS$(\pi_0|\tau)$ has a Chi-square distribution with one degree of freedom. It is easy to see that when τ is set to 1.0, the complete-case analysis is obtained, and that in that case $\tilde{\pi}_\tau = \hat{\pi}$ (i.e. the ML estimate for the complete-case analysis) and CLRS$(\pi_0|1.0) = LRS_{CC}(\pi_0)$.

D.5 Generalized Sensitivity Analysis

A generalized worst-case sensitivity analysis is to inspect for which values for the sensitivity parameter τ the lower limit of the constrained likelihood-based confidence interval meets or exceeds B. This is done by testing the conditional null hypothesis H_0: $(\pi \leq \delta|\tau)$ for successive values for τ at the one-sided 0.025 significance level.

In Table D.2, results are shown for selected values for τ. The null hypothesis H_0: $\pi \leq 0.4$ is rejected for values for τ as large as 10.0. Thus, even under the extreme assumption that the probability that the outcome of a non-seroconverted is missing is ten times as high as the probability that the outcome of a seroconverted subject is missing, the data support the conclusion that $\pi > 0.4$. Only if more extreme values for τ are assumed, the conclusion from the complete-case analysis is not supported. This can be compared with the reasons why the outcomes are missing.

Table D.2 Generalized worst-case sensitivity analysis of the example data

τ	$\tilde{\pi}_\tau$	$LL'(\tilde{\pi}_\tau, \tau)$	$LL'(0.4, \tau)$	CLRS$(0.4\|\tau)$
1.0	0.544	−62.027	−65.842	7.630
2.0	0.526	−65.140	−68.074	5.868
3.0	0.516	−66.389	−68.944	5.110
4.0	0.511	−67.062	−69.409	4.694
5.0	0.507	−67.482	−69.699	4.434
6.0	0.505	−67.769	−69.897	4.256
7.0	0.503	−67.978	−70.041	4.126
8.0	0.501	−68.137	−70.150	4.026
9.0	0.500	−68.261	−70.236	3.950
10.0	0.499	−68.361	−70.305	3.888
11.0	0.498	−68.444	−70.362	3.836
12.0	0.498	−68.513	−70.410	3.794
∞	0.490	−69.295	−70.950	3.310

In the worst-case analysis, ψ_s is assumed to be to 0.0, meaning that τ is assumed to be ∞. In that case, the log-likelihood function LL' (D.5) simplifies to

$$LL'(\pi, \tau) = s \log \pi + (n - s) \log(1 - \pi).$$

This is the log-likelihood function for the worst-case data, Form. (D.2). Thus, the log-likelihood analysis of the worst-case data yields identical results as the log-likelihood analysis with missing values τ set to ∞. This shows that the worst-case analysis is the limiting case of the generalized worst-case sensitivity analysis. The generalized analysis thus has the following nice property: if the worst-case analysis supports the complete-case analysis, so will the generalized analysis; if the worst-case analysis does not support the complete-case analysis, neither will the generalized analysis for larger values for τ.

D.6 Concluding Remarks

The advantage of the generalized worst-case sensitivity analysis is that the robustness of the complete-case analysis can be checked for less extreme assumptions than the assumption that the sensitivity parameter τ is infinite. This is a considerable gain because the assumption that τ can be infinite will rarely be realistic. Consider again the example. Suppose that four values were missing because the tube with the serum sample was broken during transport, two due to loss-to-follow-up and four because the analysis of the serum sample failed. The first reason can be assumed to be unrelated to the antibody level, but suppose that it is known that a failed serum sample analysis is more likely to occur for low antibody levels. If it is further assumed that loss-to-follow-up may also correlate with a low antibody level, then the expected number of missing positive outcomes is 2, and the expected number of missing negative outcomes 8. In that case, an estimate of the probability of a missing positive outcome is 2/43, and an estimate of the probability of a missing negative outcome is 8/57. Thus, an estimate of τ would be (8/57)/(2/43) = 3.0. For this and comparable values for τ, the generalized worst-case sensitivity analysis supported the conclusion of the complete-case analysis.

D.7 Technical Notes

The log-likelihood function (D.2) can be factorized as

$$\begin{aligned} LL(\pi, \psi, \tau) = {} & [s \log \pi + (m - s) \log(1 - \pi) + (n - m) \log(\pi/\tau + 1 - \pi)] \\ & + [s \log(1 - \psi/\tau) + (m - s) \log(1 - \psi) + (n - m) \log \psi]. \end{aligned}$$

With τ fixed, both components can be maximized independently, meaning that constraint ML estimates of ψ are independent of π. This implies that $\tilde{\psi}_{0\tau} = \tilde{\psi}_{\tau}$.

Differentiating the log-likelihood function (D.5) yields the following normal equation:

$$\frac{s}{\pi} - \frac{(m-s)}{1-\pi} + \frac{(n-m)\tau'}{1+\pi\tau'} = 0,$$

with $\tau' = (1/\tau - 1)$. Solving this equation for π produces the constrained ML estimate $\tilde{\pi}_{\tau}$. Simple algebra yields that for $\tau > 1.0$ the estimate $\tilde{\pi}_{\tau}$ is the solution to the quadratic equation

$$\pi^2(-n\tau') + \pi[(n-m+s)\tau' - m] + s = 0.$$

The roots x can be found with the quadratic formula, and $\tilde{\pi}_{\tau}$ is the root satisfying the constraint $0 < x < 1$. $\tilde{\pi}_{\tau} = s/m$ for $\tau = 1.0$ and $\tilde{\pi}_{\tau} = s/n$ for $\tau = \infty$.

Appendix E
Formula Linking Risk of Infection and Force of Infection

The formula linking the risk $\pi(t_s)$ of becoming infected during the surveillance period $[0, t_s)$ to the force of infection function $\lambda(t)$, the 'exponential formula', is [45, 46]

$$\pi(t_s) = 1 - \exp[-\int_0^{t_s} \lambda(u)du]. \tag{E.1}$$

The formula can be applied when the study population is a cohort. In this appendix, a generalization of the formula is derived that holds for both cohorts and dynamic populations in steady state.

The derivation of the generalized formula is based on the assumption that the occurrence of cases of the infectious disease is a Poisson process. A Poisson process is a counting process, used for counting the occurrences of events that appear to happen at a certain rate, but completely at random, without a certain structure. The process has the following properties:

- The members of the source population are independently subject to the force of infection.
- There can be at most one case at any time-point t after the start of the surveillance period.
- The probability of exactly one case in a small time interval $[t, t + \Delta t)$ is proportional to the length Δt of the time interval and equals $\lambda(t)\Delta t$.
- The probability of two or more cases in the interval is $o(\Delta t)$,[1] with $o(\Delta t) \to 0$ as $\Delta t \to 0$.

An alternative name for $\lambda(t)$ is *intensity of infection* function [45]. Because $\lambda(t)\Delta t$ is a probability and thus dimensionless, the dimension of $\lambda(t)$ is time^{-1}.

Let $P_0(t)$ denote the probability that there are no cases in the interval $[0, t)$. By applying the probability rule of multiplication it follows that

[1]The mathematical notation $o(\Delta t)$ represents any function of Δt which tends to 0 faster than Δt.

J. Nauta, *Statistics in Clinical and Observational Vaccine Studies*,
Springer Series in Pharmaceutical Statistics,
https://doi.org/10.1007/978-3-030-37693-2

$$P_0(t + \Delta t) = [1 - \lambda(t)\Delta t] P_0(t),$$

and

$$\frac{P_0(t + \Delta t) - P_0(t)}{\Delta t} = -\lambda(t) P_0(t).$$

This implies that

$$\frac{d P_0(t)}{dt} = -\lambda(t) P_0(t),$$

or[2]

$$\frac{P_0'(t)}{P_0(t)} = -\lambda(t).$$

Because $(\log f)' = f'/f$ it follows that

$$\frac{d \log P_0(t)}{dt} = -\lambda(t),$$

and by integrating that

$$\begin{aligned} \int_0^t -\lambda(u)du &= \int_0^t \frac{d \log P_0(u)}{du} du \\ &= \int_0^t d \log P_0(u) \\ &= \log P_0(t) - \log P_0(0) \\ &= \log P_0(t), \end{aligned}$$

because $\log P_0(0) = \log(1) = 0$. Thus

$$P_0(t) = \exp\left[-\int_0^t \lambda(u)du\right].$$

Let **T** be a non-negative continuous random variable, defined as the time until the occurrence of a case after the start of the surveillance period. Then

$$\begin{aligned} \Pr(\mathbf{T} < t_s) &= 1 - P_0(t_s) \\ &= 1 - \exp\left[\int_0^{t_s} \lambda(u)du\right]. \end{aligned} \tag{E.2}$$

Form. (E.2) is the generalization of Form. (E.1), and it holds for both cohorts and dynamic populations that are perfectly in steady state. (To see why the formula does not hold for dynamic populations that are not in steady state, consider a dynamic

[2] $P_0'(t)$ and f', shorthand notation for $d P_0(t)/dt$ and $df(t)/dt$.

population with only one member. If this member leaves the population at time-point $t < t_s$ (meaning that the population is not in steady state), the sample space of $\mathbf{T}$ is limited to $[0, t)$, unless he is replaced by a new member with the same risk profile (in which case the population is in steady state).

When the source population is a cohort, Form. (E.2) is the probability that a member becomes infected in the interval $[0, t_s)$, i.e. then $\Pr(\mathbf{T} < t_s) = \pi(t_s)$, the risk of infection during the surveillance period $[0, t_s)$, and (E.2) becomes (E.1).

Appendix F
Force of Infection Versus Hazard, Two Sides of the Same Coin

In this appendix, it is shown that although a force of infection function and a hazard function are conceptually different, they take the same functional form, and that in this sense they are synonyms.

Proof Let $\mathbf{T}$ be a non-negative random variable representing the time to the occurrence of a case of the infectious disease, after the start of the surveillance period. A hazard function $H(t)$ expresses the instantaneous probability of a case occurring at $\mathbf{T} = t$, conditional on being at risk until t. Define

$$F(t) = \Pr(\mathbf{T} \geq t).$$

The probability density function of $\mathbf{T}$ is

$$\begin{aligned} f(t) &= \lim_{\Delta t \to 0} \frac{\Pr(t \leq \mathbf{T} < t + \Delta t)}{\Delta t} \\ &= \frac{-dF(t)}{dt}. \end{aligned}$$

The hazard function is defined as [113]

$$\begin{aligned} H(t) &= \lim_{\Delta t \to 0} \frac{\Pr(t \leq \mathbf{T} < t + \Delta t \mid \mathbf{T} \geq t)}{\Delta t} \\ &= \frac{f(t)}{F(t)} \\ &= -\frac{F'(t)}{F(t)}. \end{aligned}$$

J. Nauta, *Statistics in Clinical and Observational Vaccine Studies*, Springer Series in Pharmaceutical Statistics,
https://doi.org/10.1007/978-3-030-37693-2

Because[3] $(\log f)' = f'/f$, it follows that

$$H(t) = \frac{-d \log F(t)}{dt},$$

and, by integrating, that

$$\begin{aligned}
\int_0^t H(u)du &= -\int_0^t \frac{d \log F(u)}{du} du \\
&= -\int_0^t d \log F(u) \\
&= -\log F(t) + \log F(0) \\
&= -\log F(t),
\end{aligned}$$

because $F(0) = 1$ and thus $\log F(0) = 0$. Hence

$$F(t) = 1 - \exp\left[-\int_0^t H(u)du\right],$$

and

$$\begin{aligned}
\Pr(\mathbf{T} < t) &= 1 - F(t) \\
&= 1 - \exp\left[-\int_0^t H(u)du\right].
\end{aligned} \tag{F.1}$$

From (F.1) and (E.2) it follows that

$$\exp\left[-\int_0^t H(u)du\right] = \exp\left[-\int_0^t \lambda(u)du\right], \qquad \forall\, t > 0.$$

and thus that

$$\int_0^t H(u)du = \int_0^t \lambda(u)du. \quad \forall\, t > 0.$$

Then, according to the fundamental theorem of calculus

$$H(t) = \lambda(t), \qquad \forall\, t > 0.$$

Thus, the hazard function and the force of infection function take the same functional form. □

[3] $F'(t)$ and f', shorthand notation for $dF(t)/dt$ and $df(t)/dt$.

Appendix G
Confidence Interval for the Difference of the Medians of Two Lognormal Distributions

Let $\log \mathbf{Y}_i$ be normally distributed:

$$\log \mathbf{Y}_i \sim N(\mu_i, \sigma_i^2) \qquad (i = 1, 2).$$

Then $\mathbf{Y}_i$ has a lognormal distribution with mean

$$m_i = \exp(\mu_i + \sigma_i^2/2).$$

The mean of the difference $\mathbf{Y}_1 - \mathbf{Y}_2$ is thus

$$\eta = \exp(\mu_1 + \sigma_1^2/2) - \exp(\mu_2 + \sigma_2^2/2).$$

A 95% confidence interval for η can be obtained by means of parametric bootstrapping [114]. The approach can be summarized by the following algorithm:

1. Calculate from the data the estimates $\hat{\mu}_1$, $\hat{\sigma}_1^2$, $\hat{\mu}_2$, $\hat{\sigma}_2^2$ and $\hat{\eta}$.
2. Generate N bootstrap samples (d_1, d_2), with d_i a random draw from $N(\hat{\mu}_i, \hat{\sigma}_i^2)$, and N a very large number.
3. For each bootstrap sample, calculate the bootstrap estimate for η as

$$\hat{\eta}_{\text{bootstrap}} = \exp(d_1) - \exp(d_2).$$

The N bootstrap estimates form an empirical sampling distribution for $\hat{\eta}$. A confidence interval for η can be obtained by the percentile method:

4. Find the 2.5th and the 97.5th percentile of the set of N bootstrap samples.

The medians of the two lognormal distributions are e^{μ_1} and e^{μ_2}. Assume that the parameter of interest is not η, but the difference of the two medians:

$$\omega = \exp(\mu_1) - \exp(\mu_2).$$

J. Nauta, *Statistics in Clinical and Observational Vaccine Studies*, Springer Series in Pharmaceutical Statistics,
https://doi.org/10.1007/978-3-030-37693-2

Table G.1 Reproducibility of parametric bootstrap confidence limits

	Lower limit		Upper limit	
N	1st run	2nd run	1st run	2nd run
1,000	−1.70	−1.46	8.96	10.1
10,000	−1.60	−1.70	9.74	9.58
100,000	−1.60	−1.64	9.61	9.45
1,000,000	−1.64	−1.63	9.53	9.54

Define

$$\log \mathbf{W}_i = \log \mathbf{Y}_i - \sigma_i^2/2.$$

Then

$$\log \mathbf{W}_i \sim N(\mu_i - \sigma_i^2/2, \sigma_i^2),$$

which implies that $\mathbf{W}_i$ is lognormally distributed with mean e^{μ_i}. The mean of difference $\mathbf{W}_1$–$\mathbf{W}_2$ is ω. The bootstrap approach to a confidence interval for ω is entirely analogous to the approach described above, but with the d_i draw from $N(\hat{\mu}_i - \hat{\sigma}_i^2/2, \hat{\sigma}_i^2)$.

The bootstrap method is sometimes criticized for yielding non-reproducible results. This can be avoided by setting the number of bootstrap samples N to a very large value. As an example, assume that $\hat{\mu}_1 = 1.823$, $\hat{\sigma}_1^2 = 0.169$, $\hat{\mu}_2 = 1.224$ and $\hat{\sigma}_2^2 = 0.087$. For several values for N the bootstrap analysis was run twice. The results are shown in Table G.1. With $N = 1{,}000$ the reproducibility is indeed poor, but with $N = 1{,}000{,}000$ the results are fully reproducible. On a modern laptop the total process time of the bootstrap analysis with $N = 1{,}000{,}000$ is less than 2 s.

Appendix H
SAS Code to Generate the Data Set COHORTS

SAS Code H.1

```
    data Closed_Cohort;  /*1*/
        Version="closed";
        call streaminit(4321);  /*2*/
        N0=5000;
        N1=4000;
        Ts=140;
        do F=1 to 2;
            if (F=1) then FOI="non-homogeneous";  /*3*/
            else FOI="homogeneous";
            do Group=0 to 1;
                N=N0*(Group=0) + N1*(Group=1);
                Size=N;
                Foi0=((Group=0)*1.2 + (Group=1)*0.8)/1000;  /*4*/
                Foi1=((Group=0)*2.4 + (Group=1)*1.2)/1000;
                Foi2=((Group=0)*2.0 + (Group=1)*1.0)/1000;
                Foi3=((Group=0)*1.4 + (Group=1)*0.7)/1000;
                Case=1;  /*5*/
                do Time=1 to Ts;  /*6*/
      if (F=1) then Force=Foi1*(Time<40) + Foi2*(40<=Time<80) + Foi3*(Time>= 80);
                    else Force=Foi0;
                    Ncases=rand("POISSON",N*Force);
                    do I=1 to Ncases;
                        output;
                    end;
                    N=N-Ncases;
                end;
                Case=0;  /*7*/
                Time=Ts;
                do I=1 to N;
                    output;
                end;
            end;
        end;

    data Open_Cohort;  /*8*/
        set Closed_Cohort;
        version="open";
        call streaminit(5678);  /*9*/
```


J. Nauta, *Statistics in Clinical and Observational Vaccine Studies*,
Springer Series in Pharmaceutical Statistics,
https://doi.org/10.1007/978-3-030-37693-2

```
        R=rand("UNIFORM"); /*10*/
        if (R<0.05) then
        do;
            CT=int(rand("UNIFORM")*Ts)+1;
            if (CT<Time) then
            do;
                Case=0;
                Time=CT;
            end;
        end;

    data Cohorts;
        set Closed_Cohort Open_Cohort;
        Logtime=log(Time);
        keep Version FOI Group Time Case Logtime Size;

    proc sort data=Cohorts;
        by Version FOI Group Time Case;
    run;
```

1. First create closed cohort.
2. Do not change the seed!
3. FOI: Forces of infection.
4. Forces of infection assumed in model.
5. Simulate cases and times to infection.
6. Simulate expected number of cases at t = Time.
7. Add non-cases (censored at $t = T_s$).
8. Change copy of closed cohort to open cohort by adding censoring.
9. Add censoring during surveillance period.

Appendix I
Some More SAS Codes

I.1 Wilson-Type Confidence Intervals

SAS Code I.1A Wilson-type confidence interval for a risk difference (Sect. 3.5.2)

```
data;
    Nitermax=1000;  /*1*/

/* input */

    c1=48; n1=48;  /* example 3.5 */
    c0=52; n0=52;

    confidence_level=0.95;
    precision=0.001;         /*2*/

/* calculations */

    RD=round(c1/n1-c0/n0,precision);
    n=n0+n1;
    c=c0+c1;
    critval=round(cinv(confidence_level,1),.00001); /*3*/
    do d=-1 to 1 by 2; /*4*/
        delta=RD;
        teststat=0.0;  /*5*/
        Niterations=0;
        do while ((teststat <= critval) & (Niterations < Nitermax)); /*6*/
            delta=delta+d*precision;
            L0=c0*delta*(1-delta);  /*7*/
            L1=(n0*delta-n-2*c0)*delta+c;
            L2=(n1+2*n0)*delta-n-c;
            L3=3*n;
            q=(L2/L3)**3 - L1*L2/(6*n*n) + L0/(2*n);
            p=sign(q)*sqrt((L2/(L3))**2 - L1/(L3));
            if (q=0 & p=0) then r=1;
            else r=min(q/p**3,1);
            a=(3.1416+arcos(r))/3;
            R0=2*p*cos(a)-L2/L3;
            R1=R0+delta;
```


J. Nauta, *Statistics in Clinical and Observational Vaccine Studies*,
Springer Series in Pharmaceutical Statistics,
https://doi.org/10.1007/978-3-030-37693-2

```
            vardelta=R0*(1-R0)/n0 + R1*(1-R1)/n1;
            teststat=round((((RD-delta)**2)/vardelta),.00001);
            Niterations=Niterations+1;
        end;
        CLlimit=delta-d*precision; /*8*/
        CLlimit=min(max(CLlimit,-1),1); /*9*/
        if (abs(RD)=1 & Niterations=Nitermax) then Niterations=0;
        if (d=-1) then limit="Lower"; else limit="Upper";
        output;
    end;
proc print;
        var RD limit CLlimit Niterations Nitermax precision;
run;
```

1. The reported confidence limit may not be reliable when NITERATIONS = NITERMAX which may happen when $|RD|$ is close to 1.0 or when a high precision is requested; in both cases, an increase of NITERMAX may be attempted.
2. Precision of confidence limit (i.e. number of decimals).
3. Critical value, from chi-square distribution with 1 degree of freedom.
4. d= -1: find lower confidence limit; d= $+1$: find upper confidence limit.
5. Wilson-type chi-square statistic for risk difference analysis (Z^2 Form. (3.13)).
6. Iterations: increase or decrease delta until the test statistic is > critical value.
7. Evaluate closed solutions for restricted ML estimates $\tilde{R}_0$ and $\tilde{R}_1$ (Appendix B).
8. Confidence limit is smallest or largest value for which the test statistic is $\leq$ critical value.
9. Confidence limit must fall between -1 and $+1$.

SAS Output I.1A

```
 RD  limit  CLlimit  Niterations  Nitermax  precision
0.0  Lower   -0.074       75         1000       .001
0.0  Upper    0.068       69         1000       .001
```

SAS Code I.1B Wilson-type confidence interval for a relative risk (Sect. 3.5.2)

```
data;
    Nitermax=1000;  /*1*/

/* input */

    c1=48; n1=48;  /* example 3.5 (continued) */
    c0=52; n0=52;

    confidence_level=0.95;
    precision=0.001;  /*2*/

/* calculations */

    if ((c0 > 0) & (c1 > 0)) then
```

```
        do;
            n=n0+n1;
            C=c0+c1;
            p1=c1/n1;
            p0=c0/n0;
            RR=round(p1/p0,precision);
            critval=round(cinv(confidence_level,1),.00001);  /*3*/
            do d=-1 to 1 by 2;  /*4*/
                theta=RR;
                teststat=0.0;   /*5*/
                Niterations=0;
                do while ((teststat le critval) & (Niterations < Nitermax));/*6*/
                    theta=theta+d*precision;
                    if (theta=0) then theta=theta+d*precision;
                    A=n*theta;  /7*/
                    B=-((n1+c0)*theta+n0+c1);
                    R0=(-B-sqrt(B*B-4*A*C))/(2*A);
                    R1=theta*R0;
                    vartheta=R1*(1-R1)/n1 + theta*theta*R0*(1-R0)/n0;
                    teststat=round(((p1-theta*p0)**2 / vartheta),.00001);
                    Niterations=Niterations+1;
                end;
                CLlimit=theta-d*precision;  /*8*/
                if (d=-1) then limit="Lower"; else limit="Upper";
                output;
            end;
        end;

    proc print;
        var RR limit CLlimit Niterations Nitermax precision;
    run;
```

1. The reported confidence limit may not be reliable when NITERATIONS = NITERMAX which may happen when RR is close to 0.0 or when a high precision is requested; in both cases an increase of NITERMAX may be attempted.
2. Precision of confidence limit (i.e. number of decimals).
3. Critical value, from chi-square distribution with 1 degree of freedom.
4. d= -1: find lower confidence limit; d= $+1$: find upper confidence limit.
5. Wilson-type chi-square statistic for risk ratio analysis (Z^2 Form. (3.14)).
6. Iterations: increase or decrease delta until the test statistic is $>$ critical value.
7. Evaluate closed solutions for restricted ML estimates $\tilde{R}_0$ and $\tilde{R}_1$ (Appendix B).
8. Confidence limit is smallest or largest value for which the test statistic is $\leq$ critical value.

SAS Output I.1B

```
 RR   limit  CLlimit  Niterations  Nitermax  precision
1.0   Lower   0.926        75        1000       .001
1.0   Upper   1.073        74        1000       .001
```

I.2 Barnard's Exact Test for Relative Risk Analysis

SAS Code I.2 (Sect. 3.5.3)

```
%let size=16;  /*1*/

data;
    array Z{&size,&size};
    array ci{&size};
    array cj{&size};

/* input */

    c1= 7; n1=15;   /* example 3.6 (cont.) */
    c0=12; n0=15;

    theta=0.260;  /*2*/
    sides=1;      /*3*/
     H1=">";      /*4*/

/* calculations */

    if (c1>0 & c0>0) then
    do;
        n=n1+n0;  /*5*/
    p1=c1/n1;
        p0=c0/n0;
        RR=p1/p0;
        if (c0=n0 & c1=n1) then Zobs=0;
        else
        do;
            A=n*theta;
            B=-((n1+c0)*theta+n0+c1);
            C=c0+c1;
            R0=(-B-sqrt(B*B-4*A*C))/(2*A);
            R1=theta*R0;
            vartheta=R1*(1-R1)/n1 + theta*theta*R0*(1-R0)/n0;
            Zobs=(p1-theta*p0)/sqrt(vartheta);
        end;

        do i=1 to n1; /*6*/
            do j=1 to n0;
                if (j=n0 & i=n1) then z{i+1,j+1}=0;
                else
                do;
```

```
                    p1=i/n1;
                    p0=j/n0;
                    B=-((n1+j)*theta+n0+i);
                    C=i+j;
                    R0=(-B-sqrt(B*B-4*A*C))/(2*A);
                    R1=theta*R0;
                    vartheta=R1*(1-R1)/n1 + theta*theta*R0*(1-R0)/n0;
                    Z{i+1,j+1}=(p1-theta*p0)/sqrt(vartheta);
                end;
            end;
        end;

        do i=0 to n1;  /*7*/
            if (i=0) then ci{1}=1; else ci{i+1}=ci{i}*(n1-i+1)/i;
        end;
        do j=0 to n0;
            if (j=0) then cj{1}=1; else cj{j+1}=cj{j}*(n0-j+1)/j;
        end;

        Pmax=-1; /*8*/
        do pi0=0.001 to 0.999 by 0.001;
            pi1=theta*pi0;
            if (0<= pi1 <= 1) then
            do;
                PV=0.0;
                do i=1 to n1;
                    do j=1 to n0;
                        if ((sides=1 & (Z{i+1,j+1}>=Zobs>=0 |
                            Z{i+1,j+1}<=Zobs<=0)) |
                            (sides=2 & abs(Z{i+1,j+1})>=abs(Zobs))) then
                            PV=PV + ci{i+1}*(pi1**i)*((1-pi1)**(n1-i))*
                            cj{j+1}*(pi0**j)*((1-pi0)**(n0-j));
                    end;
                end;
                if (Pmax<PV) then Pmax=PV;
            end;
        end;
        Pmax=round(Pmax,.0001);
        if (sides=1 & ((RR<theta & H1=">") | (RR>theta & H1="<"))) then
        Pmax=1-Pmax;
    end;

proc print;
    var theta sides H1 RR Zobs Pmax;
run;
```

1. Size must be (at least) $\max(n_1, n_0)+1$.
2. Null value risk ratio, theta must be > 0.
3. One- or two-sided P-value.
4. Direction of alternative hypothesis for one-sided P-values.
5. Calculate Z_{obs}.
6. Calculate the Z_{ij}.
7. Assign binomials.
8. Find maximum P-value.

SAS Output I.2

```
 theta   sides  H1       RR        Zobs       Pmax
0260       1    >   0.58333   2.35748   0.0218
```

I.3 Bootstrap Analysis to Compare Two Nelson–Aalen Risks of Infection Estimates

SAS Code I.3 (Sect. 8.2.4)

```
data Cohort;
    set Cohorts;
    where Version="open" & FOI="non-homogeneous";

proc sql;
    create table Groupsize as
    select distinct Group, Size as _Nsize_ from Cohort;
quit;

proc surveyselect data=Cohort seed=-1 out=Bootsamples
    method=pps_wr reps=10000 sampsize=Groupsize;
    size Size;
    strata Group;

data Aalen_Nelson;
    set Bootsamples;
    by Group Replicate Time;
    if first.Replicate then
    do;
        AN_est=0;
        At_risk=Size;
    end;
    if first.Time then
    do;
        D1=0;
        D2=0;
    end;
    D1 + Case*Numberhits;
    D2 + Numberhits;
    if last.Time then
    do;
        AN_est + D1/At_risk;
        At_risk + (-1)*D2;
    end;
    if last.Replicate then
    do;
        Risk_AN=1-exp(-AN_est);
        output;
    end;
```

```
    keep Group Replicate Risk_AN;

proc sort data=Aalen_Nelson;
    by Replicate Group;


proc transpose data=Aalen_Nelson out=Aalen_Nelson_trans;
    var Risk_AN;
    by Replicate;

data Aalen_Nelson_trans;
    set Aalen_Nelson_trans (rename=(Col1=R_AN0 Col2=R_AN1));
    RR=R_AN1/R_AN0;

proc univariate data=Aalen_Nelson_trans noprint;
    var RR;
    output out=out pctlpre=percentile_ pctlpts=2.5, 50.0, 97.5;

proc print data=out; run;
```

SAS Output I.3

```
percentile_  percentile_  percentile_
    2_5          50           97_5

  0.48408      0.53590      0.58907
```

I.4 Sample Sizes for Demonstrating Super Efficacy

SAS Code I.4

```
data;

/*input */

    alpha=0.025; /* one-sided significance level */
    VE=0.8;      /* example 8.4 */
    AR0=0.25;
    CSE=0.4;
    power=0.9;

    r=1;       /* randomization ratio n1/n0 */

/* calculations, see ref. [62] */

    theta0=1-CSE;
    p0=AR0;
```

```
    p1=p0*(1-VE);
    Zalpha=probit(1-alpha);
    Zbeta=probit(power);
    A=1+1/r;
    B=-(theta0*(1+p0/r)+1/r+p1);
    C=theta0*(p1+p0/r);
    R1=(-B-sqrt(B*B-4*A*C))/(2*A);
    R0=R1/theta0;
    n1=((Zalpha*sqrt(R1*(1-R1)+((theta0**2)*r)*R0*(1-R0)) +
        Zbeta*sqrt(p1*(1-p1)+((theta0**2)*r)*p0*(1-p0)) )/(p1-theta0*p0))**2;
    n0=int(n1/r)+1;
    n1=int(n1)+1;

 proc print;
    var n0 n1;
run;
```

SAS Output I.4

```
    n0      n1
   147     147
```

I.5 Sample Sizes for Comparing Two Infection Rates

SAS Code I.5

```
data;

/* input */

    VE=0.7;          /* example 8.5 */
    CSE=0.4;
    lambda0=0.05;
    alpha=0.05;
    power=0.9;
    r=2;             /* randomization ratio n1/n0 */

/* calculations */

    p=r*(1-VE)/(r*(1-VE)+1);     /* see page 112 */
    c=1;
    actual_power=0;
    do while (actual_power < power);
        c=c+1;
        c1=c;
        ready=0;
        do while (not ready);
            c1=c1-1;
            c0=c-c1;
```

```
            fu=FINV(1-alpha/2,2*(c1+1),2*c0);  /* see section 8.2.1 */
            UCL_pi=(c1+1)*fu/(c0+(c1+1)*fu);
            UCL_theta=UCL_pi/(r*(1-UCL_pi));
            LCL_ve=1-UCL_theta;
            ready=((LCL_ve>CSE)|(c1=0));
        end;
        actual_power=probbnml(pi,c,c1);
    end;
    n0=c/(lambda0*(1+r*(1-ve)));
    n1=int(r*n0)+1;
    n0=int(n0)+1;

proc print;
    var n0 n1;
run;
```

SAS Output I.5

```
    n0      n1
   1151    2302
```

References

1. Henao-Restrepo AM, Longini IM, Egger, et al. Efficacy and effectiveness of an rVSV-vectored vaccine expressing Ebola surface glycoprotein: interim results from the Guinea ring vaccination cluster-randomised trial. Lancet. 2015;386:857–66.
2. Hobson D, Curry RL, Beare AS, Ward-Gardner A. The role of serum haemagglutination-inhibiting antibody in protection against challenge infection with influenza A2 and B viruses. J Hyg. 1972;70:767–77.
3. Vesikari T, Karvonen A, Korhonen T, Espo M, Lebacq E, Forster J, Zepp F, Delem A, De Vos B. Safety and immunogenicity of RIX4414 live attenuated human rotavirus vaccine in adults, toddlers and previously uninfected infants. Vaccine. 2004;22:2836–42.
4. Harro CD, Pang YYS, Roden RBS, Hildesheim A, Wang Z, Reynolds MJ, Mast TC, Robinson R, Murphy BR, Karron RA, Dillner J, Schiller JT, Lowy DR. Safety and immunogenicity trial in adult volunteers of a human papillomavirus 16 L1 virus-like particle vaccine. JNCI. 2001;93:284–92.
5. Feiring B, Fuglesang J, Oster P, Naess LM, Helland OS, Tilman S, Rosenqvist E, Bergsaker MAR, Nøkleby H, Aaberge IS. Persisting immune responses indicating long-term protection after booster dose with meningococcal group B outer membrane vesicle vaccine. CVI. 2006;13:790–6.
6. Newcombe RG. Two-sided confidence intervals for the single proportion: comparison of seven methods. Statist Med. 1998;17:857–72.
7. Miettinen O, Nurminen M. Comparative analysis of two rates. Statist Med. 1985;4:213–26.
8. Chick SE, Barth-Jones DC, Koopman JS. Bias reduction for risk ratio and vaccine effect estimators. Statist Med. 2001;20:1609–24.
9. Suissa S, Shuster JJ. Exact unconditional sample sizes for the 2x2 binomial trial. J Royal Stat Soc, Series A. 1985;148:317–27.
10. Chan ISF. Exact tests of equivalence and efficacy with a non-zero lower bound for comparative studies. Statist Med. 1998;17:1403–13.
11. Lydersen S, Fagerland MW, Laake P. Tutorial in biostatistics: recommended tests for association in 2x2 tables. Statist Med. 2009;28:1159–75.
12. Berger RL. Multiparameter hypothesis testing and acceptance sampling. Technometrics. 1982;24:295–300.
13. Laska NS, Tang D, Meisner MJ. Testing hypothesis about an identified treatment when there are multiple endpoints. JASA. 1992;87:825–31.
14. Reed GF, Meade BD, Steinhoff MC. The reverse cumulative distribution plot: a graphic method for exploratory analysis of antibody data. Pediatrics. 1995;96:600–3.

J. Nauta, *Statistics in Clinical and Observational Vaccine Studies*,
Springer Series in Pharmaceutical Statistics,
https://doi.org/10.1007/978-3-030-37693-2

15. Small RD, Ozol-Godfrey A, Yan L. On the use of nonparametric tests for comparing immunological reverse distribution curves (RCDCs). Vaccine. 2019;37:6737–42.
16. Nauta JJP, Beyer WEP, Osterhaus ADME. On the relationship between mean antibody level, seroprotection and clinical protection from influenza. Biologicals. 2009;37:216–21.
17. Lachenbruch PA, Rida W, Kou J. Lot consistency as an equivalence problem. J Biopharm Stat. 2004;14:275–90.
18. Julious SA. Sample sizes for clinical trials. Boca Raton: Chapman & Hall/CRC; 2009.
19. Fleiss JL, Levin B, Paik MC. Statistical methods for rates and proportions. 3rd ed. Hobokon: Wiley Inc.; 2003.
20. Hirji KF, Tang ML, Vollset SE, Elashoff RM. Efficient power computation for exact and mid-P tests for the common odds ratio in several 2x2 tables. Statist Med. 1994;13:1539–49.
21. Ting Lee ML, Whitmore GA. Statistical inference for serial dilution assay data. Biometrics. 1999;55:1215–20.
22. Nauta JJP, de Bruijn IA. On the bias in HI titers and how to reduce it. Vaccine. 2006;24:6645–6.
23. Nauta JJP. Eliminating bias in the estimation of the geometric mean of HI titers. Biologicals. 2006;34:183–6.
24. Cox DR, Oakes D. Analysis of survival data. London: Chapman & Hall; 1984 (Chap. 3).
25. Wang WWB, Mehrotra DV, Chan ISF, Heyse JF. Statistical considerations for noninferiority/equivalence trials in vaccine development. J Biopharm Stat. 2006;16:429–41.
26. Schuirmann DJ. A comparison of the two one-sided tests procedure and the power approach for assessing equivalence of average bioavailability. J Pharmacokinet Biopharm. 1987;15:657–80.
27. International Conference on Harmonisation of Technical Requirements for Registration of Pharmaceuticals for Human Use. ICH harmonised tripartite guideline: statistical principles for clinical trials E9. London: European Medicines Agency for the Evaluation of Medicinal Products; 1998.
28. Joines RW, Blatter M, Abraham B, Xie F, De Clercq N, Baine Y, Reisinger KS, Kuhnen A, Parenti DL. Prospective, randomized, comparative US trial of a combination hepatitis A and B vaccine, Twinrix with corresponding monovalent vaccines. Havrix and Engerix-B in adults. Vaccine. 2001;19:4710–9.
29. Nauta J. Statistical analysis of influenza vaccine lot consistency studies. J Biopharm Stat. 2006;16:443–52.
30. Wiens BL, Iglewicz B. On testing equivalence of three populations. J Biopharm Stat. 1999;9:465–83.
31. Kong L, Kohberger RC, Koch GG. Type I error and power in noninferiority/equivalence trials with correlated multiple endpoints: an example from vaccine development trials. J Biopharm Stat. 2004;14:893–907.
32. Kong L, Kohberger RC, Koch GG. Design of vaccine equivalence/non-inferiority trials with correlated multiple binomial endpoints. J Biopharm Stat. 2006;16:555–72.
33. Committee on Proprietary Medical Products (CPMP). Points to consider on switching between superiority and non-inferiority. London: European Medicines Agency for the Evaluation of Medicinal Products; 2000.
34. Dunnett CW, Gent M. An alternative to the use of two-sided tests in clinical trials. Statist Med. 1976;15:1729–38.
35. Morikawa T, Yoshida M. A useful testing strategy in phase III trials: combined test of superiority and test of equivalence. J Biopharm Stat. 1995;5:297–306.
36. Ng T-H. Simultaneous testing of noninferiority and superiority increases the false discovery rate. J Biopharm Stat. 2007;17:259–64.
37. Leroux-Roels I, Vets E, Freese R, Seiberling M, Weber F, Salamand C, Leroux-Roels G. Seasonal influenza vaccine delivered by intradermal microinjection: a randomised controlled safety and immunogenicity trial in adults. Vaccine. 2008;26:6614–9.
38. Ganju J, Izu A, Anemona A. Sample size for equivalence trials: a case study from a vaccine lot consistency study. Statist Med. 2008;27:3743–54.

39. Kohberger RC. Comments on Sample size for equivalence trials: a case study from a vaccine lot consistency trial by J. Ganju, A Izu and A Anemona. Statist Med. 2009;28:177–8.
40. Ganju J, Izu A, Anemona A. Authors' reply. Statist Med. 2009;28:178–9.
41. Halloran ME, Longini IM, Struchiner CJ. Design and analysis of vaccine studies. New York: Springer; 2010 (Chap. 2).
42. Nelson JC, Bittner RCL, Bounds L, Zhao S, Baggs J, Donahue JG, Hambidge SJ, Jacobsen SJ, Klein NP, Naleway AL, Zangwill KM, Jackson LA. Compliance with multiple-dose vaccine schedules among older children, adolescents, and adults: results from a vaccine safety datalink study. AJPH. 2009;99:S389–97.
43. Khatib AM, Ali M, Von Seidlein L, Kim DR, Hashim R, et al. Effectiveness of an oral cholera vaccine in Zanzibar: findings from a mass vaccination campaign and observational cohort study. Lancet Infect Dis. 2012;12:837–44.
44. Ichihara MYT, Rodrigues LC, Teles Santos CAS, Teixeira MLC, De Jesus SR, Alvim De Matos SM, Gagliardi Leite JP, Mauricio L. Barretoa ML. Effectiveness of rotavirus vaccine against hospitalized rotavirus diarrhea: a case-control study. Vaccine. 2014;32:2740–7.
45. Chiang CL. Introduction to stochastic processes in biostatistics. New York: Wiley Inc.; 1968 (Sect. 3.5).
46. Miettinen O. Estimability and estimation in case-referent studies. Am J Epidemiol. 1976;103:226–36.
47. Rothman KJ, Greenland S, Lash TL. Modern epidemiology. 3rd ed. Philadelphia: Lippincott Williams & Wilkins; 2008. p. 43.
48. Rothman KJ. Epidemiology an introduction. 2nd ed. New York: Oxford University Press Inc; 2012. p. 47.
49. Nauta J, Beyer WEP, Kimp EPJA. Toward a better understanding of the relationship between influenza vaccine efficacy against specific and non-specific endpoints and vaccine efficacy against influenza infection. EBPH. 2017:14(4):e12367-1–5.
50. Jackson LA, Gaglani MJ, Keyserling HL, Balser J, Bouveret N, Fries L, Treanor JT. Safety, efficacy, and immunogenicity of an inactivated influenza vaccine in healthy adults: a randomized, placebo-controlled trial over two influenza seasons. BMC Infect Dis. 2010;10:71–5.
51. FDA. Guidance for industry: clinical data needed to support the licensure of seasonal inactivated influenza vaccines. Rockville: United States Food and Drug Administration; 2007.
52. Blennow M, Olin P, Granström M, Bernier RH. Protective efficacy of a whole cell pertussis vaccine. Br Med J. 1988;296:1570–2.
53. Aalen OO. Nonparametric inference for a family of counting processes. Ann Stat. 1978;6:701–26.
54. Nelson W. Hazard plotting for incomplete failure data. J Qual Technol. 1969;1:27–52.
55. Nelson W. Theory and applications of hazard plotting for censored failure data. Technometrics. 1972;14:945–65.
56. Borgan Ø. Nelson-Aalen estimator [Internet]. Hoboken: Wiley StatsRef: Statistics Reference Online; 2014 [cited 2018 Apr 1]. https://onlinelibrary.wiley.com/, https://doi.org/10.1002/9781118445112.stat06045.
57. Klein JP, Logan B, Harhoff M, Andersen PK. Analyzing survival curves at a fixed point in time. Statist Med. 2007;26:4505–19.
58. Chernick MR. Bootstrap methods: a guide for practitioners and researchers. 2nd ed. Hoboken: Wiley Inc; 2008 (Sect. 3.1.2).
59. Moorthy V, Reed Z, Smith PG. Measurement of malaria vaccine efficacy in phase III trials: report of a WHO consultation. Vaccine. 2007;25:5115–23.
60. Jahn-Eimermacher A, Du Prel JB, Schmitt HJ. Assessing vaccine efficacy for the prevention of acute otitis media by pneumococcal vaccination in children: a methodological overview of statistical practice in randomized controlled clinical trials. Vaccine. 2007;25:6237–44.
61. Fleming TR, Harrington DP. Counting processes and survival analysis. New York: Wiley; 1991.
62. Farrington CP, Manning G. Test statistics and sample size formulae for comparative binomial trials with null hypothesis of non-zero risk difference or non-unity relative risk. Statist Med. 1990;9:1447–54.

63. Greenland S, Pearl J, Robins JM. Causal diagrams for epidemiologic research. Epidemiology. 1999;10:37–48.
64. Pearl J, Glymour M, Jewell NP. Causal inference in statistics: a primer. New York: Wiley; 2016.
65. VanderWeele TJ. On the distinction between interaction and effect modification. Epidemiology. 2009;20:863–71.
66. Greenland S, Thomas DC. On the need for the rare disease assumption in case-control studies. Am J Epidemiol. 1982;116:547–53.
67. Rothman KJ, Greenland S, Lash TL. Modern epidemiology. 3rd ed. Philadelphia: Lippincott Williams & Wilkins; 2008. p. 250–2.
68. Jackson ML, Nelson JC. The test-negative design for estimating influenza vaccine effectiveness. Vaccine. 2013;31:2165–8.
69. Sullivan SG, Cowling BJ. "Crude vaccine effectiveness" is a misleading term in test-negative studies of influenza vaccine effectiveness. Epidemiology. 2015;26:e60.
70. Sato T. Estimation of a common risk ratio in stratified case-cohort studies. Stat Med. 1992;11:1599–605.
71. Sato T. Maximum likelihood estimation of the risk ratio in case-cohort studies. Biometrics. 1992;48:1215–21.
72. Jackson ML, Chung JR, Jackson LA, et al. Influenza vaccine effectiveness in the united states during the 2015–2016 season. N Engl J Med. 2017;377:534–43.
73. Miettinen OS. Theoretical epidemiology: principles of occurrence research in medicine. New York: Wiley Inc; 1985. p. 227–9.
74. Royston P, Sauerbrei W. Multivariable model-building: a pragmatic approach to regression analysis based on fractional polynomials for modelling continuous variables New York: Wiley Inc; 2008.
75. Harrell FE Jr. Regression modelling strategies: with applications to linear models, logistic and ordinal regression, and survival analysis. 2nd ed. Cham: Springer; 2015.
76. Groenwold RH, Klungel OH, Altman DG, van der Graaf Y, Hoes AW, Moons KG. Adjustment for continuous confounders: an example of how to prevent residual confounding. CMAJ. 2013;185(5):401–6.
77. Rothman KJ, Greenland S, Lash TL. Modern epidemiology. 3rd ed. Philadelphia: Lippincott Williams & Wilkins; 2008.
78. Austin PC. An introduction to propensity score methods for reducing the effects of confounding in observational studies. Multivariate Behav Res. 2011;46(3):399–424.
79. Simpson C, Lone N, Kavanagh K, Ritchie L, Robertson C, Sheikh A, et al. Seasonal influenza vaccine effectiveness (SIVE): an observational retrospective cohort study–exploitation of a unique community-based national-linked database to determine the effectiveness of the seasonal trivalent influenza vaccine. HS&DR. 2013;1(10).
80. Sutton AJ, Abrams R, Jones DR, Sheldon TA, Song F. Methods for meta-analysis in medical research. Chichester: Wiley Ltd; 2000.
81. DerSimonian R, Laird N. Meta-analysis in clinical trials. Controlled Clin Trials. 1986;7:177–88.
82. Yusuf S, Peto R, Lewis J, Collins R, Sleight P. Beta blockade during and after myocardial infarction: an overview of the randomized trials. Prog Cardiovasc Dis. 1985;27:335–71.
83. De Oliveira LH, Camacho LAB, Coutinho ESF, Ruiz-Matusa C, Leite JPG. Rotavirus vaccine effectiveness in Latin American and Caribbean countries: a systematic review and meta-analysis. Vaccine. 2015;33(Suppl 1):A248–54.
84. Van Houwelingen HC, Arends LR, Stijnen T. Advanced methods in meta-analysis: multivariate approach and meta-regression. Statist Med. 2002;21:589–624.
85. Darvishian M, Bijlsma MJ, Hak E, Van den Heuvel ER. Effectiveness of seasonal influenza vaccine in community-dwelling elderly people: a meta-analysis of test-negative design case-control studies. Lancet infect Dis. 2014;14:1228–39.
86. Colditz GA, Brewer T, Berkey C, Wilson M, Burdick E, Fineberg H, Mosteller F. Efficacy of BCG vaccine in the prevention of tuberculosis: meta-analysis of the published literature. JAMA. 1994;271:698–702.

87. Berkey CS, Hoaglin DC, Mosteller F, Colditz GA. A random-effects regression model for meta-analysis. Statist Med. 1995;14:395–411.
88. World Health Organization. Correlates of vaccine-induced protection: methods and implications. Geneva: WHO Press; 2013.
89. Plotkin SA, Gilbert PB. Nomenclature for immune correlates of protection after vaccination. Clin Infect Dis. 2012;54:1615–7.
90. Qin L, Gilbert PB, Corey L, McElrath MJ, Self SG. A framework for assessing immunological correlates of protection in vaccine trials. J Infect Dis. 2007;196:1304–12.
91. Prentice RL. Surrogate endpoints in clinical trials: definition and operational criteria. Statist Med. 1989;8:431–40.
92. Dunning AJ. A model for immunological correlates of protection. Statist Med. 2006;25:1485–97.
93. Forrest BD, Pride MW, Dunning AJ, Capeding MR, Chotpitayasunondh T, Tam JS, Rappaport R, Eldridge JH, Gruber WC. Correlation of cellular immune responses with protection against culture-confirmed influenza virus in young children. Clin Vaccine Immunol. 2008;15:1042–53.
94. White CJ, Kuter BJ, Ngai A, Hildebrand CS, Isganitis KL, Patterson CM, Capra A, Miller WJ, Krah DL, Provost PJ, Ellis RW, Calandra GB. Modified cases of chickenpox after varicella vaccination: correlation of protection with antibody response. Pediatr Infect Dis J. 1992;11:19–23.
95. Jódar L, Butler J, Carlone G, Dagan R, Goldblatt D, Käyhty H, Klugman K, Plikaytis B, Siber G, Kohberger R, Chang I, Cherian T. Serological criteria for evaluation and licensure of new pneumococcal conjugate vaccine formulations for use in infants. Vaccine. 2003;21:3265–72.
96. Thompson WW, Price C, Goodson B, et al. Early thimerosal exposure and neuropsychological outcomes at 7 to 10 years. N Engl J Med. 2007;357:1281–92.
97. Heron J, Golding J, ALSPAC Study Team. Thimerosal exposure in infants and developmental disorders: a prospective cohort study in the United Kingdom does not support a causal association. Pediatrics. 2004;114:577–83.
98. Madsen KM, Lauritsen MB, Pedersen CB, et al. Thimerosal and the occurrence of autism: negative ecological evidence from Danish population-based data. Pediatrics. 2004;112:604–6.
99. Andrews N, Miller E, Grant A, Stowe J, Osborne V, Taylor B. Thimerosal exposure in infants and developmental disorders: a retrospective cohort study in the United kingdom does not support a causal association. Pediatrics. 2004;114:584–91.
100. Ellenberg SS. Safety considerations for new vaccine development. Pharmacoepidemiol Drug Saf. 2001;10:411–5.
101. Mutsch M, Zhou W, Rhodes P, Bopp M, Chen RT, Linder T, Spyr C, Steffen R. Use of the inactivated intranasal influenza vaccine and the risk of Bell's palsy in Switzerland. NEJM. 2004;350:896–903.
102. Farrington P, Whitaker H, Weldeselassie YG. Self-controlled case series studies: a modelling Guide with R. New York: Chapman & Hall/CRC; 2018.
103. Dmitrienko A, Tamhane AC, Bretz F. Multiple testing problems in pharmaceutical statistics. New York: Chapman & Hall/CRC; 2010.
104. Benjamini Y, Hochberg Y. Controlling the false discovery rate: a practical and powerful approach to multiple testing. J Royal Stat Soc, Series B. 1995;57:289–300.
105. Mehrotra DV, Heyse JF. Use of the false discovery rate for evaluating clinical safety data. Stat Methods Med Res. 2004;13:227–38.
106. Berry SM, Berry DA. Accounting for multiplicities in assessing drug safety: a three-level hierarchical mixture model. Biometrics. 2004;60:418–26.
107. Singh M, editor. Vaccine adjuvants and delivery systems. Hoboken: Wiley Inc; 2007.
108. FDA. Guidance for industry: toxicity grading scale for healthy adults and adolescent volunteers enrolled in preventive vaccine clinical trials. Rockville: United States Food and Drug Administration; 2007.
109. De Bruijn IA, Nauta J, Gerez L, Palache AM. The virosomal influenza vaccine Invivac: immunogenicity and tolerability compared to an adjuvanted influenza vaccine, Fluad in elderly subjects. Vaccine. 2006;24:6629–31.

110. Lehmann EHJM. Nonparametrics methods based on ranks. New York: Springer Science+ Business Media; 2006.
111. Diggle L, Deeks J. Effect of needle length on incidence of local reactions to routine immunisation in infants aged 4 months: randomised controlled trial. Br Med J. 2000;321:931–3.
112. FDA. Guidance for industry: clinical data needed to support the licensure of pandemic inactivated influenza vaccines. Rockville: United States Food and Drug Administration; 2007.
113. Kalbfleisch JD, Prentice RL. The statistical analysis of failure time data. 2nd ed. New York: Wiley Inc; 2002 (Sect. 1.2.1).
114. Chen YH, Zhou XH. Interval estimates for the ratio and difference of two lognormal means. Statist Med. 2006;25:4099–113.

Index

J. Nauta, *Statistics in Clinical and Observational Vaccine Studies*,
Springer Series in Pharmaceutical Statistics,
https://doi.org/10.1007/978-3-030-37693-2

Printed in the United States
by Baker & Taylor Publisher Services